Chemie, Physik, und Technologie der Kunststoffe
in Einzeldarstellungen
Herausgegeben von K. A. Wolf

—— 12 ——

O. Griffin Lewis

Physical Constants
of Linear Homopolymers

Springer-Verlag Berlin · Heidelberg · New York 1968

Dr. O. Griffin Lewis
American Cyanamid Company
Polymer Chemistry Section
Chemical Department
Stamford, Conn. 06904 — U.S.A.

Title No. 4312

ISBN-13: 978-3-642-46114-9   e-ISBN-13: 978-3-642-46112-5
DOI: 10.1007/ 978-3-642-46112-5

Preface

This book is a compilation of some fundamental properties of polymers, arranged alphabetically in one table. It should prove useful in practical applications of polymers and in the development of theories of polymer behavior. Much of the impetus for studies of new polymers derives from the desire to understand how molecular structure influences physical properties. A large quantity of data has been generated in pursuit of this goal, and certain consistent trends have been discovered. It is hoped that further progress in this area will be accelerated by bringing together the published data on polymers in this form.

The physical properties listed were selected for the following reasons. Firstly, they are fundamental quantities from which other properties may be deduced. Secondly, they can be determined reproducibly in different laboratories, and finally, they have been reported for a sufficient number of polymers to justify inclusion.

The table of polymers has been arranged so as to economize on space, to keep the size of the book within reasonable proportions, and to facilitate scanning. The device of inverting the names tends to group similar compounds to aid in searching. Because of the ease with which the table can be surveyed it was felt that supplementary indexes by melting point or glass temperature would be unnecessary. References are given to the literature cited, but unfortunately it is not possible to indicate the reference next to each datum without greatly expanding the size of the table.

The need for a systematic method of storage and retrieval of information on such a large number of compounds is obvious, but the absence of a widely accepted system of nomenclature for polymers is a serious impediment to the use of an alphabetical index.

I am pleased to acknowledge my indebtedness to Mrs. CHRISTEL KAPPES WATSON for invaluable assistance in developing the system, described herein, which makes alphabetizing practical, and to Miss BARBARA ALLSTROM for help with problems of naming and indexing.

It was on the advice and encouragement of Dr. WALTER M. THOMAS that the decision was made to publish the material as a book. Professor MAURICE MORTON was kind enough to bring the compilation to the attention of the Committee on Macromolecular Chemistry of the National Academy of Sciences – National Research Council, whose suggestion it was to contact the Springer Publishing Company. It is a pleasure to acknowledge my debt to many friends at the American Cyanamid Company for assistance rendered and to the Company for permission to publish.

Special thanks are due to Dr. EDWARD V. THOMPSON, Dr. LAWRENCE V. GALLACHER, Dr. MARY L. MILLER and Professor KARL A. WOLF for many helpful criticisms of the manuscript.

Stamford, February 1968   O. GRIFFIN LEWIS

Contents

List of Symbols

b (after number)	Brittle point
d_a	Density of amorphous phase
d_c	Density of crystalline phase
$d\bar{v}/dT$	Temperature coefficient of specific volume
D (subscript)	Wavelength of sodium D line
E	Internal energy
k	Boltzmann's constant
n	Refractive index
N_1	Number of solvent molecules
N_2	Number of polymer molecules
s (after number)	Softening point
T_d	Temperature of dynamic loss peak
T_g	Glass temperature or other second-order transition
T_m	Crystalline melting point or other first-order transition
T_s	Temperature of equilibrium between crystalline polymer and solution of same polymer
V	Total volume
V_1	Molar volume of solvent
V_2	Molar volume of polymer
V_u	Volume per mole of polymer repeating units
w_{11}	Interaction energy between nearest neighbor solvent molecules
w_{12}	Interaction energy between solvent and polymer
w_{22}	Interaction energy between nearest neighbor polymer segments
x	Number of segments per polymer chain
z	Lattice coordination number
α, β, γ (after number)	Crystalline polymorphs
ΔA_m	Helmholtz energy of mixing
ΔE_m	Internal energy of mixing
ΔG_u	Change in Gibbs energy per repeat unit
ΔH_u	Enthalpy of fusion per repeat unit
ΔS_m	Entropy of mixing
$\varnothing_1$	Volume fraction of solvent
$\varnothing_2$	Volume fraction of polymer
χ	Polymer-solvent interaction constant
I, II, III (after number)	Crystalline polymorphs

Introduction

Polymers are composed of giant molecules formed from a limited number of different types of structural units connected in long chains. The chains may have branches at intervals along their lengths and they may be linked together at different points. The physical properties of polymers depend upon the distribution of structural units, branches and crosslinks, yet these structural features are difficult to determine experimentally. At the present time the only polymers whose structures can be characterized precisely are linear homopolymers, that is polymers comprised of a single basic unit connected to two, and only two, other similar units. Only linear homopolymers are listed in this compilation. Copolymers which are known by their method of preparation to have a regularly alternating structure are also included, since they conform with the above definition of a homopolymer. The structural unit is composed of two parts in this case, and it is only because of the use of source names that they are named copolymers rather than homopolymers. Random copolymers are not listed.

Unreasonably strict application of the definition of a linear homopolymer would relegate it to hypothetical materials only, since structural irregularities are inevitable in polymer synthesis. In practice, it merely means that the expected structural unit is bifunctional, and that branches (trifunctional units) or crosslinks (units of higher functionality) are believed to be relatively infrequent. The word homopolymer includes all polymers prepared from a single monomer (or pair of co-reagents where two compounds, such as a diol and dibasic acid, must be condensed to form the polymer). The resulting structure may be a random sequence of isomeric units. For example, polymers of conjugated dienes contain repeating units formed by 1,2 addition, as well as 1,4 addition. The 1,2 units possess stereoisomers and the 1,4 units have geometrical isomers. The relative proportions of the different structural units depend upon polymerization conditions, so that it is necessary either to designate the known structure as determined by nuclear magnetic resonance, infrared spectroscopy or X-ray diffraction, or to specify those polymerization conditions which regulate the structure. In the absence of such designation or specification the polymer may be understood to be the normal random mixture of isomers made without stereoregulating catalysts.

The properties selected for inclusion are not greatly dependent on polymer molecular weight provided that this exceeds some critical value, usually in the range 30,000 to 100,000. Reported values have therefore been listed only if the report gives some evidence that the polymer was of reasonably high molecular weight.

The physical properties tabulated are discussed in some detail in the following pages under separate headings. The purpose of the discussion is to acquaint the reader with the meaning and uses of these values, to give some indication of the ways of judging their reliability, and to explain the reasons for selecting some data in preference to others. Only the underlying principles are discussed. Experimental procedures are far from uniform in this field and it would be out of place to attempt to describe all of them here. The reader can usually find a description of the experimental method in the original reference, and in fact the table provides a means of searching for references to procedures.

References are collected in the last column, listed in the order in which the data quoted appear in the table, reading from left to right. Additional references not quoted are given in parentheses where space permits.

Solubility Parameter

The effects which low molecular weight liquids exert on polymers are of great practical and theoretical interest. On the one hand a method is required for predicting plasticizer efficiencies, solvent power, solution viscosity, melting-point depression and the swelling of gels. On the other hand measurements of these quantities can be made to yield information concerning the intermolcular forces which influence other physical properties of polymers. A useful parameter for this purpose is the square root of the cohesive energy density, which HILDEBRAND[1] has designated the solubility parameter. The cohesive energy density of a liquid is the energy of vaporization per unit volume.

Cohesive energy densities are calculated from calorimetric heats of vaporization or from vapor pressure data taken over a range of temperature. Solubility parameters for various liquids have been reported by HILDEBRAND and SCOTT[1]; BURRELL[2]; CROWLEY, TEAGUE and LOWE[3]; WALKER[4]; and BRISTOW and WATSON[5].

High polymers have negligibly small vapor pressure, so the cohesive energy density must be estimated indirectly, making use of solubility theory. The most widely used theory of polymer solutions is that developed by FLORY[6] and HUGGINS[7]. Since many of the heats of fusion as well as solubility parameters reported here were derived by application of this theory, the nature of the theory is discussed briefly below.

The Flory-Huggins theory is a generalization of the BRAGG-WILLIAMS[8] approximation in the lattice model of binary solutions. The polymer is considered to consist of x segments equal in size to a solvent molecule. Hence x is the ratio of molar volumes of the polymer and solvent. N_2 polymer molecules and N_1 solvent molecules are placed randomly on a lattice of coordination number z. The volume fractions of solvent and polymer are then

$$\emptyset_1 = N_1/(N_1 + xN_2) \tag{1a}$$

$$\emptyset_2 = N_2/(N_1 + xN_2) \tag{1b}$$

The entropy of mixing is evaluated by considering the number of possible configurations or arrangements of the N_1, N_2 molecules on the $N_1 + xN_2$ lattice sites, relative to the number when there are no solvent molecules. The result is

$$\Delta S_m/k = -N_1 \ln \emptyset_1 - N_2 \ln \emptyset_2 \tag{2}$$

The energy of mixing is calculated by means of a variation of the quasichemical method. Only nearest-neighbor interactions are considered, and in a binary system only

[1] HILDEBRAND, J. H., and R. L. SCOTT: The Solubility of Nonelectrolytes. New York: Reinhold 1950.
[2] BURRELL, H.: Interchem. Rev. *14*, 3 (1955); Offic. Dig. Federation Soc. Paint Technol. *27*, 726 (1955).
[3] CROWLEY, J. D., G. S. TEAGUE, Jr., and J. W. LOWE, Jr.: J. Paint Technol. *38*, 269 (1966).
[4] WALKER, E. E.: J. Appl. Chem. (London) *2*, 470 (1952).
[5] BRISTOW, G. M., and W. F. WATSON: Trans. Far. Soc. *54*, 1731 (1958).
[6] FLORY, P. J.: Principles of Polymer Chemistry. New York, Ithaca: Cornell University Press 1953.
[7] HUGGINS, M. L.: Ann. N. Y. Acad. Sci. *41*, 1 (1942); J. Am. Chem. Soc. *64*, 1712 (1942); Ind. Eng. Chem. *35*, 216 (1943).
[8] BRAGG, W. L., and E. J. WILLIAMS: Proc. Roy. Soc. (London) *A145*, 699 (1934).

three types of nearest-neighbor interactions occur, whose energies are designated w_{11}, w_{22} and w_{12}. Since the probability that any site is occupied by a solvent molecule is $\emptyset_1$, and by a polymer segment $\emptyset_2$, it follows that the total energy of the system is

$$-E = \frac{z}{2} (N_1 + xN_2) (\emptyset_1^2 w_{11} + \emptyset_2^2 w_{22} + 2\emptyset_1 \emptyset_2 w_{12}) \qquad (3)$$

The energies of the components before mixing are

$$-E_1 = \frac{z}{2} N_1 w_{11} \qquad (4a)$$

$$-E_2 = \frac{z}{2} xN_2 w_{22} \qquad (4b)$$

The energy of mixing is the difference between these quantities, or

$$-\Delta E_m = \frac{z}{2} (N_1 + xN_2) \emptyset_1 \emptyset_2 (w_{11} + w_{22} - 2w_{12}) \qquad (5)$$

This equation is then recast in the form

$$\Delta E_m = kT \chi N_1 \emptyset_2 \qquad (6)$$

where

$$\chi = -\frac{z}{2} (w_{11} + w_{22} - 2w_{12}) / kT \qquad (7)$$

is a dimensionless quantity which characterizes the net interaction energy per solvent molecule divided by kT. The Helmholtz energy of mixing is obtained by combining equations (2) and (6):

$$\Delta A_m / kT = \Delta E_m / kT - \Delta S_m / k = N_1 \ln \emptyset_1 + N_2 \ln \emptyset_2 + \chi N_1 \emptyset_2 \qquad (8)$$

When $2w_{12} > w_{11} + w_{22}$, χ and the energy of mixing are both positive. From criteria for thermodynamic stability it can be shown[6] that a critical point for phase separation should occur at

$$\chi_c = (1 + x^{1/2})^2 / 2x \qquad (9)$$

and

$$\emptyset_{2c} = 1 / (1 + x^{1/2}) \qquad (10)$$

As $x \to \infty$, $\chi_c \to 1/2$ and $\emptyset_{2c} \to 1/x^{1/2}$. If $\chi < 1/2$, the polymer and solvent are miscible in all proportions.

It should be noted that the relations for Helmholtz energy of mixing and the critical point contain no lattice parameters. Nevertheless the use of a single lattice to characterize both polymer and solvent is an artifice which must detract from the generality of the theory. Furthermore the random mixing approximation must fail near the critical point, and the nearest-neighbor approximation ignores long-range interactions which may well be important. In spite of these objections the theory has been found to be in satisfactory semi-quantitative agreement with many experimental observations and there is no doubt that it contains the essential features which determine the effects exerted by low molecular weight liquids on polymers.

In the Scatchard-Hildebrand[1] treatment of this problem the pair potential w_{12} is related to the properties of the pure components by making the geometric mean approximation:

$$w_{12} = (w_{11} w_{22})^{1/2} \qquad (11)$$

This approximation is valid when the interaction arises predominantly from London dispersion forces. Then, with the cohesive energy density defined as

$$E_1/V_1 = -zw_{11}/2 V_1 \qquad (12)$$

where V_1 is the molar volume, Equation (5) can be written

$$\Delta E_m = V \varnothing_1 \varnothing_2 \left[\left(\frac{E_1}{V_1} \right)^{1/2} - \left(\frac{E_2}{V_2} \right)^{1/2} \right]^2 \qquad (13)$$

where V is the total volume of the system. The quantity $(E/V)^{1/2}$ is defined as the solubility parameter δ. The theory predicts that $\Delta E_m = 0$ when $\delta_1 = \delta_2$, and that $\Delta E_m > 0$ for $\delta_1 \neq \delta_2$. Consequently the best solvents are liquids whose solubility parameters are similar to that of the polymer.

Due to hydrogen bonding and other specific interactions between solvent and solute, for which the geometric mean relation is inappropriate, Equation (13) does not predict the energy of mixing very accurately. For non-polar systems, however, the solubility parameter is a fairly reliable index of relative solvent power. The values listed in the table were generally obtained by plotting some property such as intrinsic viscosity of a soluble polymer, or degree of swelling of a cross-linked polymer against solubility parameter of the solvents. The maximum viscosity or swelling is assumed to occur at $\delta_2 = \delta_1$. A less reliable technique is to determine the solubility parameter range for which solubility is complete and assume that δ_2 is at the midpoint of this range.

Melting Point

The term melting point is defined thermodynamically as the temperature at which the crystalline and liquid phases of a pure substance are in equilibrium at a pressure of one atmosphere. The conditions implicit in this definition are very difficult to meet in the determination of melting points of high polymers. Even the most careful fractionation does not yield a polymer completely uniform in molecular weight and completely regular in structure. A pure crystalline phase is equally difficult to obtain. The growth of a polymer crystal from the melt is greatly inhibited by the high viscosity or low coefficient of self-diffusion which is characteristic of polymers. The formation of a pure crystal phase would require selection of molecules of identical length by the growing crystal front, disentanglement of these from the liquid, and ordering in perfect register with the extended-chain molecules already in the crystal. This is not likely to occur except under very unusual conditions. Ordinarily it is found that little or no selection according to molecular weight takes place, but that the crystals accommodate molecules of any length by a process of chain folding, with the chain ends either excluded from the crystal or incorporated as defects. Folded-chain crystals are metastable with respect to the extended-chain crystal because of the contribution of the folds to the free energy of the system. Annealing or treatment with swelling liquids results in extensive rearrangement and recrystallization, an increase in melting point and decrease in melting range.

Almost perfect extended-chain crystals of certain polymers have been grown in sizes large enough to exhibit no size dependence of the thermodynamic properties. Such crystals have a pronounced tendency to superheat, and extremely low heating rates must be used to measure the melting point. If the polymers were of high molecular weight and highly regular in structure, the melting point on slow heating was accepted as the preferred value for that polymer. Samples containing ordinary folded-chain crystals melt over a wide range due to the presence of crystals of varying size and of perfection. The upper limit of the melting range is listed in the table, and in general the highest melting point was chosen from among those reported.

Polymorphism is quite common in polymers. Wherever reported, the observed upper temperature limit of each form is listed, in descending order, using the author's designations for the different forms. If no designations have been given the transition temperatures are labeled with Roman numerals, in order of decreasing temperature.

The melting point appears on a differential thermal analysis (DTA) curve as an endothermic peak, on a specific heat curve as a slow rise in specific heat with increasing temperature followed by a sharp drop, in dilatometry as a rapid increase in volume with increasing temperature, which terminates abruptly at the liquidus curve, and in X-ray diffraction as the disappearance of sharp diffraction lines from the diagram. All physical properties are quite markedly time dependent just below the melting point due to the metastable nature of semi-crystalline polymers. Hence the observed melting points is time dependent, higher values being found at lower heating rates.

Melting points can be determined microscopically as the temperature at which the last trace of crystalline birefringence disappears. This should not be confused with the birefringence which is due to the orientation of amorphous material, and which disappears just above the glass temperature. Capillary melting points are frequently reported for polymers, but these are not reliable if the degree of crystallinity is low or the melt is too viscous to permit rapid fusion of the particles. Capillary melting points are listed only if better data are not available, and only if the polymers are highly crystalline and melt sharply.

Softening points or penetrometer measurements are frequently used in studying the melting of polymers. Such values are listed as melting points, but with the symbol s, and only if more reliable values have not been reported. The polymer must first be demonstrated to be highly crystalline, and the softening must be abrupt, as expected for a solid-liquid transition.

The polymer melt temperature (PMT), or temperature at which the sample begins to leave a trail when rubbed along a hot bar, does not seem to be a true melting point, and these values are not listed.

Heat of Fusion

The melting of a crystalline substance is a first-order transition, accompanied by an isothermal increase in enthalpy, the latent heat of fusion. Due to the impossibility of obtaining a pure crystalline phase in polymers, melting occurs over a range, and the enthalpy-temperature curve is S-shaped rather than discontinuous. To calculate the heat of fusion from calorimetric measurements entails extrapolation of the curve for the crystalline form to the melting point, which reduces the accuracy of the measurement. Since the quantity required is the heat of melting one mole of crystalline units, the calorimetric heat must be corrected for the fraction of amorphous polymer present, and this introduces a further uncertainty in the determination.

An alternative method which avoids these problems is to apply the Flory[6] theory of melting point depression of polymer solutions. A crystalline polymer may be in equilibrium with a solution of the same polymer at a temperature T_s, lower than the melting point of the pure polymer T_m. At equilibrium the chemical potentials of the polymer in the two phases must be equal, i.e. $\mu_2^c = \mu_2(\emptyset_2)$. This can be expressed as

$$\mu_2^c - \mu_2^\circ = \mu_2(\emptyset_2) - \mu_2^\circ \tag{14}$$

where μ_2° is the chemical potential per mole of repeating units in the standard state, which we take to be the pure liquid polymer at the same temperature and pressure. The left hand side of Equation (14) is just the negative of the Gibbs energy of fusion ΔG_u. This can be written

$$\mu_2^c - \mu_2^\circ = -\Delta G_u = -\Delta H_u (1 - T_s/T_m) \tag{15}$$

The right hand side of Equation (14) is the chemical potential of the polymer in solution relative to that in the standard state, which may be obtained by differentiating the Helmholtz energy of mixing, as given by Equation (8), with respect to the number of polymer repeating units. Note that the number of repeating units is given by $x N_2 V_1/V_u$, where V_u is the volume per mole of repeating units. Hence

$$\mu_2(\emptyset_2) - \mu_2^\circ = \frac{V_u}{x V_1} \left(\frac{\partial \Delta A_m}{\partial N_2} \right)_{V, T, N_1} = RT(V_u/V_1) \lfloor x^{-1} \ln \emptyset_2 - (1 - x^{-1}) \emptyset_1 + \chi \emptyset_1^2 \rfloor \tag{16}$$

For large x

$$\mu_2(\emptyset_2) - \mu_2^\circ \cong RT(V_u/V_1)(\chi \emptyset_1^2 - \emptyset_1) \tag{17}$$

Substituting Equations (15) and (17) in Equation (14) and rearranging, we obtain

$$1/T_s - 1/T_m = (R/\Delta H_u)(V_u/V_1)(\emptyset_1 - \chi \emptyset_1^2) \tag{18}$$

The experimental melting points of a series of mixtures can be plotted as $(1/T_s - 1/T_m)/\emptyset_1$ against $\emptyset_1$, whence the intercept equals $RV_u/\Delta H_u V_1$ and the slope equals this quantity multiplied by χ. Heats of fusion determined in this manner are generally consistent with the calorimetric values and both are listed in the table.

There seems to be no objective way of choosing a best value when those reported are in disagreement, so all those reported are listed. In those cases where the number of observations is large, a few representative of the range are given.

Heats of solid-solid transitions must be determined directly, either by calorimetry or by measurement of the area under the endothermic peak on a differential-thermal analysis trace. The heats of transition from one polymorphic crystal form to another are listed in order of decreasing temperature, each value having the designation of the form which is stable below the transition temperature.

Density

The density of an amorphous polymer may be determined by conventional methods appropriate for the physical form of the sample: pycnometry, application of Archimedes' principle, or measurement of position in a density gradient column.

The density of a crystalline phase must be determined indirectly in polymers. Examination of crystalline polymers by X-ray diffraction shows periodicity analogous to that found in simple crystals, but also reveals a background of diffuse scattering arising from the presence of disordered material. Electron and light microscopy frequently show amorphous material between the crystalline aggregates. The preferred method of determination of the crystalline density in the presence of this amorphous phase is by analysis of the X-ray diffraction pattern.

Polymer chains crystallize in the form of helices of uniform pitch, the planar zig-zag configuration being considered a helix with one asymmetric unit per turn. The helices are parallel and equidistant and usually lie in fixed relative positions with respect to translation along and rotation about the helical axis. The asymmetric unit of a polymer consists of either one monomer repeat unit, or in the case of syndiotactic polymers two monomer repeat units, rather than the entire molecule. In this regard polymer crystals differ from those of low molecular weight compounds. The crystal lattice is a multiple of a basic element, the unit cell. The unit cell dimensions, the number of chains passing through the unit cell, and the length of the asymmetric unit can be deduced by analysis of the X-ray diffraction pattern of a crystalline, preferably oriented, sample. From this information the density of the pure crystalline phase can be calculated. Chain ends do not normally form an ordered array in high polymers, but are rejected from the crystal phase and accommodated in some fashion as defects. The unit cell dimensions from which the crystalline density is calculated actually refer to a sub-cell containing no chain ends.

Given the densities of the completely amorphous and completely crystalline forms of a polymer, the degree of crystallinity can be computed in a straightforward way by assuming additivity of volumes. Conversely, if the degree of crystallinity is known from X-ray or infrared measurements on samples of widely different densities, the crystalline and amorphous densities can be calculated. Some of the values in the table were calculated in this way, where for example completely amorphous samples could not be produced.

The simple two-phase model seems to be inappropriate for certain polymers, and crystallinities calculated from density measurements may be in poor agreement with

X-ray, infrared or n.m.r. determinations. However, the latter are often in disagreement among themselves. The density method is most likely to fail under conditions where different polymorphic phases may coexist, where paracrystalline phases of varying degrees of perfection are possible, and below the glass temperature, where the amorphous density may be dependent on crystallinity and rate of cooling.

The densities of amorphous and crystalline forms are listed in the table as d_a and d_c, respectively, with the temperature of measurement given as a superscript. If there is no superscript the temperature was not reported, but was probably about 20 or 25 °C. Densities enclosed in parentheses are values determined directly on semicrystalline samples and are listed under d_c if no published value was found for the crystalline density, or under d_a if no value is known for the amorphous density. The highest semicrystalline density is entered under d_c, the lowest under d_a. In addition, if a direct determination was higher than the calculated d_c, both are listed. This is an indication that the X-ray diffraction analysis may be in error. Multiple determinations of d_a and d_c are included for completeness. Densities reported for crystalline polymorphs are indicated by whatever symbol (Greek letter or Roman numeral) was used in the original reference. If no designation was given the polymorphs are numbered by Roman numerals in order of decreasing transition temperature.

Glass Temperature

If the temperature of a supercooled liquid can be reduced far enough below the normal freezing point without crystallization occurring, the specific heat and thermal expansivity may be observed to decrease abruptly over a narrow range of temperature. Below this temperature range the material is in a vitreous or glassy state and possesses many of the physical properties of solids while retaining the amorphous X-ray diffraction pattern of a liquid. Vitrification has been observed in all types of liquids and is an important aspect of the thermal behavior of polymers.

The structure of liquids can be described as a state of short-range quasicrystalline order. At low temperatures the positions and relative orientations of nearest neighboring molecules are fairly well defined, although not as precisely as in a crystal. As the temperature is raised the average number of nearest neighbors and the extent of short range order continuously decrease due to the increasing rate of molecular motion. Accompanying these structural changes there are energy and volume changes which contribute to the observed specific heat and thermal expansion. Structural changes are greatly inhibited in the glassy state and the corresponding contributions to specific heat and thermal expansion are therefore absent.

A change in the position or orientation or conformation of any one molecule in a liquid inevitably affects all of its neighbors. Thus any kind of molecular motion results in a temporary disruption of the local order. The lower the temperature and the greater the degree of order, the greater must be the degree of cooperation in molecular motions. This accounts for the large positive entropies and enthalpies of activation which are observed for viscous flow and mechanical and dielectric relaxation in liquids at low temperatures, and for the fact that activation entropies and enthalpies increase as the temperature drops. The changes in short-range order and number of nearest neighbors

8

which accompany a temperature change require molecular movement, and since a similar disruption of the local order must accompany this motion it must also have a large activation entropy and enthalpy. Below a certain temperature molecular motion becomes too slow to permit establishment of the equilibrium liquid structure in experiments of normal duration, resulting in the observed decrease in specific heat and thermal expansivity. The transformation of a liquid to a glass is therefore the result of a relaxation phenomenon and the measured transition temperature depends on the time allowed for equilibration. In order to establish uniform practice in reporting and comparing transition temperatures, KAUZMANN[9] suggested that the glass temperature be defined as "that temperature at which the specific heat or the thermal expansion coefficient of the liquid shows a more or less sudden change due to relaxation effects in experiments allowing something like 10 minutes to 1 hour for equilibrium to be reached." The glass temperatures listed in the table correspond to this definition except for those determined by differential thermal analysis (DTA), which may be slightly high due to the relatively high heating rate used.

The range of temperature over which the transformation takes place is the difference between the temperature at which the relaxation process just becomes noticeable during a measurement and the temperature at which the relaxation time is negligible. This range is generally about 10 to 20°. Below the transformation range the glass is metastable and its properties quite reproducible. The densities and expansivities of glassy polymers are listed in the table of physical constants.

Softening points of polymers are frequently reported instead of glass temperatures. This is the temperature at which the compliance of the specimen reaches a stated value under standardized conditions of stress and heating rate. The measurement depends upon viscoelastic properties which vary from one material to another, and the softening point may be either higher or lower than the glass temperature. A study[10] of some standard softening point tests indicated that the Vicat softening point and the B.S. 1493 cantilever bending test gave good agreement with dilatometric glass temperatures of a series of substituted polystyrenes. However, even the best softening point tests can be greatly in error (for determination of the glass temperature) for crosslinked or crystalline polymers, or those such as polyacrylonitrile which undergo relatively little softening at the glass temperature. Softening points are listed in the table in the Tg column with the designation "s", but only if the glass temperature has not been reported for that polymer.

Similarly the brittle point is listed, designated "b", if no glass temperature has been reported. This is the temperature at which a specimen of standard dimensions breaks under standard impact conditions.

The "polymer melt temperature" (PMT) is the temperature at which a specimen begins to leave a trail when drawn along a hot bar with a temperature gradient. This method also depends primarily on the viscoelastic behavior of the sample but has not been correlated with known transitions. These values are not included in the table.

Dilatometric measurements on certain polymers show several abrupt changes in thermal coefficient of expansion. Presumably each of these singularities results from a relaxation effect associated with a different type of structural change. For the purpose of concise description of the thermal behavior of polymers they can be considered to be transitions, even if proof of this is lacking. All transitions of this type are listed in the Tg column in order of decreasing temperature.

9 KAUZMANN, W.: Chem. Revs. *43*, 219 (1948).
10 BARB, W. G.: J. Polymer Sci. *37*, 515 (1959).

The glass transition can be observed in crystalline as well as amorphous polymers, although the transition temperature and magnitude of the change in specific heat or expansivity are usually dependent on the degree of crystallinity. Nevertheless these transitions are treated in the same way as for amorphous polymers.

The glass temperature is generally found to increase with molecular weight of the polymer, approaching asymptotically to a limiting value at high molecular weight. The limiting value is attained for all practical purposes in the range of molecular weights from about 30,000 in the case of certain highly polar polymers to 100,000 for non-polar compounds.

None of the reported values were obtained by empirical methods such as extrapolation to higher molecular weight or by the use of estimation schemes. The transitions were those actually observed on real samples. In view of the fact that the most likely source of error in determining glass temperatures is low molecular weight or the presence of impurities such as solvents, moisture or unreacted monomer, and that these invariably depress the glass temperature, the highest value reported was selected for inclusion in the table. Exceptions to this rule were made only when there was reason to believe the higher value was spurious. References to other values are given in parentheses.

Damping Peak Temperature

It was shown in the preceding section that the glass transition is the result of a relaxation process and that the molecular motion involved is closely related to the motions responsible for viscous flow and mechanical or dielectric relaxation. It follows from this that rheological or dielectric measurements can be used to detect transitions in polymers. In particular the temperature of maximum mechanical or dielectric loss corresponds to the center of a transition range which varies with the frequency of the measurement.

The rigidity of a polymer is a function of time due to the relief of stresses by molecular rearrangement. A sudden deformation (change in shape or volume) perturbs the equilibrium (or metastable) structure of the system and raises its free energy. The system returns to a state of lower free energy by a slow diffusion process, accompanied by relaxation of the stress. Several types of relaxation processes may be distinguished in polymers. Long-range changes in chain conformation, independent rearrangements of short segments of the chain, and reorientation of pendant side groups all contribute in different degrees and at different rates to the relaxation.

Since molecular motion in polymers is an activated process the effect of temperature is to change the rates of all the motions contributing to relaxation. In and above the glass transition range relaxation rates change very rapidly with temperature due to the large activation energy in this temperature region. Measurements carried out at constant frequency or constant time after the initial deformation reveal a pronounced drop in stiffness and a gradual change from glassy response at low temperature to rubbery response above the transition range. The most convenient method of studying this behavior is by means of cyclic deformations.

When a polymer is subjected to a sinusoidally varying stress of a given frequency, the strain will also vary sinusoidally but will lag behind the stress by a certain amount.

The stress per unit strain may be resolved into two components, one in phase with strain and the other 90° out of phase. The in-phase component is known as the storage modulus and represents the elastic response of the sample. The out-of-phase component is known as the loss modulus and is proportional to the energy dissipated as heat per cycle.

Although the above argument has been phrased in terms of rheological properties, it can be applied also to dielectric properties. Polarization by an electric field is analogous to shear strain, the dielectric constant is analogous to the storage modulus, and the dielectric loss factor is analogous to the loss modulus. The two kinds of experiment differ in sensitivity to motions of various types because of differences in the interactions of molecules with the different force fields. However, mechanical and dielectric relaxation rates are of similar magnitude and have virtually identical activation energies.

Measurements of dielectric and mechanical loss as a function of frequency have shown that relaxation is not a simple exponential process in polymers and hence cannot be expressed in terms of a single relaxation time. The dispersion in dielectric constant and storage modulus is spread over a wide frequency range and the loss curves exhibit broad maxima. The dispersion regions in the frequency spectrum are directly related to the transition phenomena discussed in the preceding section. The reason is that the molecular motions required for changes in the liquid configuration are quite similar to those responsible for dielectric and mechanical relaxation.

The various types of relaxation processes in polymers give rise to several peaks in the curve of dielectric or mechanical loss as a function of temperature. At low frequency the peaks are sharpest, best resolved and closest to the underlying transitions in dilatometric or calorimetric measurements. For this reason only the peak temperatures at the lowest reported frequency were tabulated.

It is a peculiar feature of energy dissipation measurements that the nature of the transition responsible for the dispersion (first or second order, amorphous or crystalline phase) cannot be deduced without making supplementary tests. While in many cases this was done, the conclusions are not always unequivocal, and so the values are simply listed in descending order with no attempt to specify the order of the transition or the phase responsible.

The sensitivity of a dynamic test to the occurrence of a transition or onset of molecular motion depends upon the magnitude of the transition and the coupling with the force field. That is, the motion must be active in absorbing whatever form of energy is applied, in order to be observed. The table does not indicate whether the values listed were obtained by dielectric or by mechanical measurements. This is not to say that different methods are equivalent, but merely that whatever test was used was sufficiently sensitive to detect the transition.

Thermal Expansion Coefficient

The bond lengths between segments of a polymer chain are virtually independent of temperature and impose restraints on the expansion which occurs as the temperature is raised. Polymer systems therefore have lower expansivities than related liquids which are not polymeric. Below the glass temperature the expansivity is reduced still further because of the absence of structural changes which contribute to expansion in liquids.

The expansivity of a polymer is not exactly independent of temperature but generally shows a gradual increase with increasing temperature, perceptible only if the temperature range covered is large. It is convenient to ignore the gradual increase compared with the sudden jump in expansivity on passing through a transition, and to represent the volume-temperature curve by two straight lines intersecting at the transition point. Along with glass temperatures determined in this way it is common practice to report the observed values of $d\bar{v}/dT$ immediately above and below the transition. These are entered in the table, separated with a slash, the glassy coefficient given first. Values enclosed in parentheses represent determinations on partly crystalline samples and are reported only if no published values for completely amorphous samples have been found.

Most of the thermal expansion coefficients were determined by volume dilatometry. Measurement of the buoyancy in a liquid whose density is accurately known as a function of temperature is also an accurate method. Some values were determined from refractive index data at different temperatures, assuming the molar refraction to be a constant. Values reported as $\bar{v}^{-1}d\bar{v}/dT$ were converted to $d\bar{v}/dT$ where necessary by the use of available density data. Linear coefficients of expansion were simply multiplied by three to obtain $\bar{v}^{-1}d\bar{v}/dT$, then converted to $d\bar{v}/dT$ using reported densities.

Multiple transitions have been observed on occasion by dilatometry, and where the expansion coefficients were reported for the respective temperature ranges these are listed on a line with the transition temperature, in order of decreasing temperature.

Nomenclature and Indexing

An index should be usable with a minimum of effort on the reader's part. Searching can be facilitated to some extent by proper choice of indexing procedure and arrangement of the table but it is not possible to anticipate all potential interests and needs of the readers. The primary consideration in design of the table was to assign each compound unambiguously to its position in the file so that all data and references are collected in one place, and so that the user will either find the polymer of interest directly or know that it is not listed.

Classification of polymers by functional type is not generally suitable because some of the classes would be excessively large and the reader might have to search several classes to find a particular compound. Indexing by formula requires too much time and effort in searching for common polymers. The best solution is a simple alphabetical listing. To make an alphabetical listing practical it is necessary to define a set of rules of nomenclature.

Polymer molecules which differ only in length, end-groups, or in the frequency of sone occasional structural irregularity such as branching, but which are composed predominantly of identical repeating units, are ordinarily considered to be the same compound. Hence it is only necessary to name the repeating unit. Systematic names for the diradicals which comprise the repeating units in polymers are awkward and unfamiliar to most chemists. This type of nomenclature was proposed[11,12] by the

[11] J. Polymer Sci. 8, 257 (1952).
[12] HUGGINS, M. L., G. NATTA, V. DESREUX, and H. MARK: Makromol. Chem. 82, 1 (1965); J. Polymer Sci. 56, 153 (1962).

Subcommission on Nomenclature of the IUPAC Commission on Macromolecules but has not come into general use.

In this compilation common usage is more nearly followed by naming the source materials, or monomers, and the name of the functional group, if any, incorporated into the chain backbone. Since the same polymer can often be formed in a variety of ways, from different source materials, it is necessary to specify arbitrarily the class of source materials from which each functional type of polymer shall be named. For example polyesters shall be named according to the dibasic acid and diol from which they could have been formed, even though some other route to the polymer may actually have been used. They are indexed only under the name of the acid. The source materials chosen for naming the various types of polymers are tabulated below.

Polymer Type	Source Material	Structure
Polyacetals	Aldehyde, polyacetal	$$\underset{\displaystyle -\overset{\textstyle R}{C}H-O-}{}$$
	Aldehyde, polyacetal with diol	$$-R'-O-\overset{\textstyle R}{C}H-O-$$
Polyamides	Acid, amino-, polyamide	$$-R-\overset{\textstyle O}{\overset{\|}{C}}-NH-$$
	Acid, polyamide with diamine	$$-R-\overset{O}{\overset{\|}{C}}-NH-R'-NH-\overset{O}{\overset{\|}{C}}-$$
Polyanhydrides	Acid, polyanhydride	$$-R-\overset{O}{\overset{\|}{C}}-O-\overset{O}{\overset{\|}{C}}-$$
Polyanhydrideamides	Acid, polyanhydride	$$-O-\overset{O}{\overset{\|}{C}}-R-\overset{O}{\overset{\|}{C}}-NH-R-NH-\overset{O}{\overset{\|}{C}}-R-\overset{O}{\overset{\|}{C}}-$$
Polycarbonates	Carbonic acid, polyester with diol	$$-R-O-\overset{O}{\overset{\|}{C}}-O-$$
Polyesters	Acid, hydroxy-, polyester	$$-R-\overset{O}{\overset{\|}{C}}-O-$$
	Acid, polyester with diol	$$-R-\overset{O}{\overset{\|}{C}}-O-R'-O-\overset{O}{\overset{\|}{C}}-$$
Polyesteramides	Highest-ranking acid, polyester with diol	$$-R-O-\overset{O}{\overset{\|}{C}}-R'-\overset{O}{\overset{\|}{C}}-NH-R''-NH-\overset{O}{\overset{\|}{C}}-R'-\overset{O}{\overset{\|}{C}}-O-$$
Polyesterurethanes	Highest-ranking acid, polyester with diol	$$-R-O-\overset{O}{\overset{\|}{C}}-\overset{R'}{\overset{\|}{N}}-R''-\overset{R'}{\overset{\|}{N}}-\overset{O}{\overset{\|}{C}}-O-R-O-\overset{O}{\overset{\|}{C}}-R'''-\overset{O}{\overset{\|}{C}}-O-$$
Polyethers	Cyclic ether, polyether	$$-R-O-$$
	Diol, polyether	$$-R-O-$$
	Diol, polyether with bivalent radical	$$-R-O-R'-O-$$
	See also Polyacetal	

Polymer Type	Source Material	Structure
Polyimides	Acid, polyimide with diamine	$-R-N\begin{smallmatrix}C=O\\C=O\end{smallmatrix}R'\begin{smallmatrix}O=C\\O=C\end{smallmatrix}N-$
Polyketones	Ketene, derivative, polyketone	$-\underset{R'}{\overset{R}{C}}-\overset{O}{C}-$
Polyketones	Unsaturated monomer, polyketone (*alt*) with carbon monoxide	$-CH_2-CH-\overset{R}{\underset{}{\overset{O=}{C}}}-$
Polymercaptals	Aldehyde, thio-, polymercaptal	$-\underset{R}{CH}-S-$
Polymercaptals	Aldehyde, polymercaptal with dithiol	$-R'-S-\underset{R}{CH}-S-$
Polymercaptoles	Ketone, polymercaptole with dithiol	$-R-S-\underset{R''}{\overset{R'}{C}}-S-$
Polyphosphazenes	Phosphonitrile, derivative, polyphosphazene	$-\underset{R}{\overset{R'}{P}}=N-$
Poly(selenoacetals)	Aldehyde, seleno-, poly(selenoacetal)	$-\underset{R'}{CH}-Se-$
Polysiloxanes	Silicone, derivative, polysiloxane	$-\underset{R}{\overset{R'}{Si}}-O-$
Polysiloxanes	Silanol, bivalent radical-, polysiloxane	$-\underset{R}{\overset{R'}{Si}}-R''-\underset{R}{\overset{R'}{Si}}-O-$
Polysulfones	Cyclic sulfone, polysulfone	$-R-\underset{O}{\overset{O=}{S}}=O-$

Polymer Type	Source Material	Structure
Poly(thioesters)	Acid, mercapto-, poly(thioester)	$-R-\overset{\overset{\displaystyle O}{\parallel}}{C}-S-$
Poly(thioesters)	Acid, poly(thioester) with dithiol	$R'-\overset{\overset{\displaystyle O}{\parallel}}{C}-S-R-S-\overset{\overset{\displaystyle O}{\parallel}}{C}-$
Poly(thioethers)	Cyclic sulfide, poly(thioether)	$-R-S-$
	See also Polymercaptals and Polymercaptoles	
Polyureas	Carbonic acid, polyurea with diamine	$-R-NH-\overset{\overset{\displaystyle O}{\parallel}}{C}-NH-$
Polyureas	Carbonic acid, polyurea (*alt*) with diamines 1 and 2	$-NH-R_1-NH-\overset{\overset{\displaystyle O}{\parallel}}{C}-NH-R_2-NH-\overset{\overset{\displaystyle O}{\parallel}}{C}-$
Polyurethans	Isocyanic acid, hydroxyalkyl ester, polyurethan	$-R-NH-\overset{\overset{\displaystyle O}{\parallel}}{C}-O-$
Polyurethans	Isocyanic acid, bivalent radical ester, polyurethan with diol	$-R-O-\overset{\overset{\displaystyle O}{\parallel}}{C}-NH-R'-NH-\overset{\overset{\displaystyle O}{\parallel}}{C}-O-$
Polyurethans	N-Carboxylic acid, hydroxy-, polyester*	$-O-\overset{\overset{\displaystyle O}{\parallel}}{C}-\overset{\overset{\displaystyle R}{\mid}}{N}-R'-$
Polyurethans	N-Carboxylic acid, polyester with diol*	$-R-O-\overset{\overset{\displaystyle R'}{\mid}}{C}-N-R''-\overset{\overset{\displaystyle O}{\parallel}}{N}-C-O-$
Polyvinyls	Unsaturated monomer	$-\overset{\overset{\displaystyle R\ \ R''}{\mid\ \ \ \mid}}{\underset{\underset{\displaystyle R'\ R'''}{\mid\ \ \ \mid}}{C-C}}-$
Other Carbon-Chain Polymers	Cyclic monomer	$-CH_2-CH_2-\overset{\overset{\displaystyle R'}{\mid}}{\underset{\underset{\displaystyle R}{\mid}}{C}}-$

*Only when N atom is completely substituted.

16

Many polymers are named after real or hypothetical cyclic compounds as source materials as seen above. However, in some cases, especially where rings occur in the polymer chain, such names would be ambiguous or unnecessarily complex. These polymers are named from the parent ring compound and a bivalent radical which alternates with it (designated by *alt*) and which modifies the functional name. As an example ring-opening polymerization of norbornene gives the polymer

named cyclopentane, 1,3-*trans*-vinylene polymer *(alt)*. Polyethers of bisphenols are named in a similar way. The compound

is named phenol, 4,4'-isopropylidenebis[2,6-dichloro-, 2-hydroxytrimethylene diether with 4,4'-isopropylidenediphenol, 2-hydroxytrimethylene polyether *(ar, ar', alt)*.

The polymer name may consist of as many as four parts: (1) the primary source material, under which the polymer is indexed; (2) the functional type, that is, the type of functionality incorporated into the backbone; (3) the co-reagent; and (4) the structure designation. The structure designation is necessary because more than one mode of polymer formation may be possible from the same source materials. Structural isomers, geometrical isomers, and stereoisomers must be differentiated, as well as structures which are not isomeric. The different polymer structures which may be formed in the polymerizations of conjugated dienes are designated as shown in the case of 1,3-butadiene:

1,3-Butadiene, *cis*-1,4-polymer

– – –, *trans*-1,4-polymer

– – –, 1,2-polymer

Polymerization by a free radical mechanism results in a copolymer of units of all three types. In the absence of any structure designation the polymer may be understood to be the free radical product.

In principle it is possible for cyclic or unsaturated monomers to form either head-to-head, tail-to-tail polymers; or head-to-tail polymers. In practice nearly all polymers whose structures have been determined have been found to be characterized by a high degree of head-to-tail regularity. This structure is therefore assumed to predominate in all polymers of this type and it is unnecessary to add the designation "head-to-tail" to the polymer name.

A somewhat analogous type of structural isomerism occurs in certain polymers formed by condensation of two co-reagents, for example the polyesters of 1,2-propanediol with dibasic acids, or polyamides of 3-methylhexanedioic acid. Two different structural units are possible:

$$-O-CH_2-\underset{\underset{\displaystyle CH_3}{|}}{CH}-O-\overset{\overset{\displaystyle O}{||}}{C}-R-\overset{\overset{\displaystyle O}{||}}{C}- \qquad \text{or} \qquad -O-\underset{\underset{\displaystyle CH_3}{|}}{CH}-CH_2-O-\overset{\overset{\displaystyle O}{||}}{C}-R-\overset{\overset{\displaystyle O}{||}}{C}-$$

and

$$-NH-R-NH-\overset{\overset{\displaystyle O}{||}}{C}-CH_2-CH_2-\underset{\underset{\displaystyle CH_3}{|}}{CH}-CH_2-\overset{\overset{\displaystyle O}{||}}{C}-, \qquad \text{or} \qquad -NH-R-NH-\overset{\overset{\displaystyle O}{||}}{C}-CH_2-\underset{\underset{\displaystyle CH_3}{|}}{CH}-CH_2-CH_2-\overset{\overset{\displaystyle O}{||}}{C}-$$

respectively. The resulting polymers are of course random copolymers of the two units. Again, this need not be designated in the polymer name.

If a chain carbon atom (or other tetrahedral atom) in a high polymer bears two different lateral substituents, inversion of the disposition of the substituents results in a different stereoisomer which is not enantiomorphous with the original compound. Farina, Peraldo and Natta proposed[13] the use of the term diasteric to describe such atoms, independent of their asymmetry, since in general they give rise to diastereoisomers. A tactic polymer is defined[14] as one in which there exists an orderly relationship between the structures of all the diasteric centers in the chains, whereas an atactic polymer is sterically disordered. If a hypothetical observer advancing along the bonds constituting the main chain finds all the diasteric atoms to have similar configurations when viewed along the preceding chain bond, the chain is said to be isotactic. If the hypothetical observer finds a regular alternation in configuration the chain is said to be syndiotactic.

If the repeating unit of the chain contains two distinguishable diasteric atoms, five different tactic structures can be envisioned :

1. monoisotactic
2. monosyndiotactic
3. *erythro*-diisotactic
4. *threo*-diisotactic
5. disyndiotactic.

The terms isotactic and syndiotactic refer as above to the structure of each diasteric atom relative to comparable atoms in the other repeating units. The prefix *erythro* indicates that when a pair of adjacent diasteric atoms is rotated into an eclipsed conformation, at least two similar substituents can be superimposed. *Threo* denotes the nonsuperimposable isomer. In a disyndiotactic polymer each diasteric center is *erythro* to one of its neighbors and *threo* to the other, so there is only one isomer.

The tacticity is indicated in parentheses following the name of the functional type of polymer. If the polymer is presumed because of its crystallinity to be tactic, but the configuration has not yet been established, it is simply designated stereoregular. In the absence of any indication of tacticity the polymer may be understood to be atactic.

Having reduced the problem of polymer nomenclature to one of source nomenclature, it is now necessary to choose one of the established systems of naming organic compounds. Chemical Abstracts has long set the standard for systematic chemical nomenclature for indexing purposes, which at the same time conforms to accepted

[13] FARINA, M., M. PERALDO, and G. NATTA: Angew. Chem., Intern. Ed. Eng. *4*, 107 (1965).
[14] NATTA, G., M. FARINA, and M. PERALDO: J. Polymer Sci. *43* (1960).

chemical usage, and this system is adopted here. The Chemical Abstracts system is periodically brought up-to-date, and the rules used here are detailed in the "Introduction to the Subject Index" of Volume 56 (January-June 1962).

In deference to certain practices which are almost universal in naming polymers, some exceptions have been made to the Chemical Abstracts' rules. These are listed below.

1. Acrylic acid derivatives with one substituent in the 2 position are listed under Acrylic acid, except for Methacrylic acid.

2. Acrylonitrile derivatives with one substituent in the 2 position are listed under Acrylonitrile, except for Methacrylonitrile.

3. Cyclic ethers with rings of more than five atoms are named Oxide, (poly)-methylene.

4. Cyclic sulfides with rings of more than five atoms are named Sulfide, (poly)-methylene.

5. Cyclic sulfones are names Sulfone, (poly)methylene; or Sulfone, phenylethylene; etc.

If the reader is unfamiliar with the Chemical Abstracts' rules of nomenclature it is still possible to find the desired entry by use of the Formula Index to Chemical Abstracts as a guide. However, it was not intended that the reader should have to study a set of rules or consult a formula index in order to use the table. Rather, it is hoped that any chemist can locate an entry even with no previous familiarity with the system, but perhaps with a little perseverance. Cross-references to alternative names are included as an aid to the reader. The practice of name inversion used herein should facilitate searching, since derivatives of the same parent are thereby grouped together.

The list of polymers is also a list of monomers, or source materials. However, certain compounds (diols, diamines, dithiols) might never appear as primary indexing names, but can be found by searching under all possible primary sources with which they might react to form a polymer.

As a result of the conventions used in forming the polymer names some of the source materials named are hypothetical compounds or monomers not suitable (by reason of instability or preference for ring formation) for direct conversion to the polymer. The original references must be consulted for the actual monomers used.

Although random copolymers are excluded from this listing, some copolymers are known from their method of preparation to have a regular alternating structure. These are listed with the names of the comonomers separated by the designation *alt*, for alternating. However, where more than one functional group is incorporated into the chain backbone of an alternating copolycondensate a separate system of naming must be used. See the table of defined source materials for polyesteramides and polyester-urethans as examples.

Table of Polymers

See page VII for meaning of symbols

No.	Polymer	$\dfrac{\delta}{\text{cal}^{1/2}\,\text{cm}^{-3/2}}$	$\dfrac{\varDelta H_u}{\text{cal mol}^{-1}}$	n
A 1	**Acenaphthylene**			
A 2	**Acetaldehyde,** polyacetal			
A 3	−, polyacetal (isotactic)			
A 4	−, polymercaptal with 1,10-decanedithiol			
A 5	−, polymercaptal with 1,6-hexanedithiol			
A 6	**Acetic acid,** vinyl ester. See **Vinyl acetate.**			
A 7	−, *p*-vinylphenyl ester			
A 8	−, **amino**-. See **Glycine.**			
A 9	*p, p'*- **biphenylylenedi**-, polyester with ethylene glycol			
A 10	−, **chloro**-, vinyl ester			1.513_D^{25}
A 11	−, **cyclohexyl**-. See **Cyclohexaneacetic acid.**			
A 12	−, **(ethylenedioxy)di**-, polyamide with ethylenediamine			
A 13	−, −, polyamide with 1,6-hexanediamine			
A 14	−, −, polyamide with piperazine			
A 15	−, **(ethylenedithio)di**-, polyanhydride			
A 16	−, **hydroxy**-. See **Glycolic acid.**			
A 17	−, **[isopropylidenebis(*p*-phenyleneoxy)]di**-, polyanhydride			
A 18	−, **oxydi**-. See **Diglycolic acid.**			
A 19	−, **[*p*-phenylenebis(methylenethio)]di**-, polyanhydride			
A 20	−, *p*-**phenylenedi**-. See *p*-**Benzenediacetic acid.**			
A 21	−, **(*m*-phenylenedioxy)di**-, polyanhydride			
A 22	−, **(*p*-phenylenedioxy)di**-, polyanhydride			
A 23	−, **thiol**-, vinyl ester			
A 24	−, **trifluoro**-, vinyl ester		(1800)	1.375_D^{25}
A 25	**Acetone,** polyacetal			
A 26	**Acetylene**			
A 27	**Acrylamide**			
A 28	−, *N-sec*-**butyl**-			
A 29	−, *N-tert*-**butyl**-			
A 30	−, *N, N*-**dibutyl**-			

No.	$\dfrac{10^4\, d\bar{v}/dT}{\text{cc g}^{-1}\text{ deg}^{-1}}$	$\dfrac{d_a}{\text{g cm}^{-3}}$	$\dfrac{d_c}{\text{g cm}^{-3}}$	$\dfrac{T_g}{°C}$	$\dfrac{T_d}{°C}$	$\dfrac{T_m}{°C}$	References
A 1				214			320, (96), (43)
A 2	2.1/6.3	1.071			$-29_{0.58}$		546, 265, 404
					-100_1		404
A 3			1.14			165	80, 63, 405
A 4						35	241
A 5						60	241
A 6							
A 7				83s			199
A 8							
A 9						150	180
A 10	1.3/3.4	1.45		31			351, 326, 191, 199
A 11							(14)
A 12						168	148
A 13						126	148
A 14						165	148
A 15						83	229, 239
A 16							
A 17						202	229
A 18							
A 19						88	229, 239
A 20							
A 21						130	229
A 22						152	229
A 23				60s			199
A 24				46		175	204, 359, 126, (199)
A 25			1.23			60	218
A 26		(1.10)	1.15				483
A 27		1.302^{23}		165			93
A 28				117			93
A 29				128			93
A 30				60s			109

No.	Polymer	$\dfrac{\delta}{\text{cal}^{1/2}\,\text{cm}^{-3/2}}$	$\dfrac{\varDelta H_u}{\text{cal mol}^{-1}}$	n
A 31	**Acrylamide, *N,N*-dibutyl-,** (isotactic)			
A 32	—, *N, N*-**diisopropyl**-			
A 33	—, —, (stereoregular)			
A 34	—, *N, N*-**dimethyl**-			
A 35	—, —, (stereoregular)			
A 36	—, *N*-**dodecyl**-			
A 37	—, *N*-**isopropyl**-	10.7		
A 38	—, —, (stereoregular)			
A 39	—, *N*-**methyl**-			
A 40	—, *N*-**(1-methylbutyl)**-			
A 41	—, *N*-**(1-methylheptyl)**			
A 42	—, *N*-**(1-methylnonyl)**-			
A 43	—, *N*-**(1-methyloctyl)**-			
A 44	—, *N*-**(1-methylpentyl)**-			
A 45	—, *N*-**methyl**-*N*-**phenyl**-			
A 46	—, —, (stereoregular)			
A 47	**Acrylic acid**			
A 48	—, benzyl ester			
A 49	—, *p*-(butoxycarbonyl)phenyl ester			
A 50	—, butyl ester	8.7		1.466_D^{20}
		8.80		
		9.05		
A 51	—, —, (isotactic)			
A 52	—, *sec*-butyl ester			
A 53	—, —, (isotactic)			
A 54	—, *tert*-butyl ester			
A 55	—, —, (isotactic)			
A 56	—, 3-chloro-2,2-bis(chloromethyl)propyl ester			

No.	$\dfrac{10^4\,d\bar{v}/dT}{\text{cc g}^{-1}\text{ deg}^{-1}}$	$\dfrac{d_a}{\text{g cm}^{-3}}$	$\dfrac{d_c}{\text{g cm}^{-3}}$	$\dfrac{T_g}{°C}$	$\dfrac{T_d}{°C}$	$\dfrac{T_m}{°C}$	References
A 31		(0.98)	1.06			350s	484, 109
A 32				120s			109
A 33				110		350s	109
A 34	1.6/4.6						460, (109)
	1.7/						460
A 35						300s	109
A 36				47			93
A 37		1.028^{23}		85			305, 93,
		1.07					127
A 38			(1.118)			200	127
A 39				105s			199
A 40				107			93
A 41				66			93
A 42				40			93
A 43				52			93
A 44				71			93
A 45				180s			109
A 46						330s	109
A 47				106			90, (56)
A 48	1.5/4.1			6			460
A 49	2.9/5.5			13			460
A 50	2.6/6.0	1.087^{26}		−49	-40_1		234,326,2,190,93,113
					-110_1		339,97,328,(16),(235)
					-190_1		(62)
A 51						47	217
A 52	2.9/6.5			−17			460, (235), (318)
A 53		(1.05)	1.06	−23		130	217, 318
A 54				43			318
A 55		1.00^{25}	1.04	40		200	53, 217, 318
			$(1.078)^{25}$				53
A 56	1.5/3.9			46			460

No.	Polymer	δ $\mathrm{cal^{1/2}\,cm^{-3/2}}$	ΔH_u $\mathrm{cal\,mol^{-1}}$	n
A 57	**Acrylic acid,** 2-chloroethyl ester			
A 58	—, *o*-chlorophenyl ester			
A 59	—, *p*-cyanobenzyl ester			
A 60	—, 2-(2-cyanoethoxy)ethyl ester			
A 61	—, 2-cyanoethyl ester			
A 62	—, 2-(2-cyanoethylthio)ethyl ester			
A 63	—, 3-(2-cyanoethylthio)propyl ester			
A 64	—, cyanomethyl ester			
A 65	—, 2-(cyanomethylthio)ethyl ester			
A 66	—, 6-(cyanomethylthio)hexyl ester			
A 67	—, *p*-cyanophenyl ester			
A 68	—, 2-(3-cyanopropylthio)ethyl ester			
A 69	—, cyclohexyl ester			
A 70	—, 2,4-dichlorophenyl ester			
A 71	—, *m*-(dimethylamino)phenyl ester			
A 72	—, 1,3-dimethylbutyl ester			
A 73	—, 2,2-dimethylpropyl ester			
A 74	—, *m*-(ethoxycarbonyl)phenyl ester			
A 75	—, *o*-(ethoxycarbonyl)phenyl ester			
A 76	—, *p*-(ethoxycarbonyl)phenyl ester			
A 77	—, 2-ethoxyethyl ester			1.471_D^{25}
A 78	—, 3-ethoxypropyl ester			1.465_D^{25}
A 79	—, ethyl ester	9.2		1.4685_D^{20}
		9.35		
		9.4		
A 80	—, 2-ethylbutyl ester			
A 81	—, 2-ethylhexyl ester			
A 82	—, 4-ethyl-l-isobutyloctyl ester			
A 83	—, 1-ethylpropyl ester			
A 84	—, 2-(ethylthio)ethyl ester			

No.	$\dfrac{10^4 \, d\bar{v}/dT}{\text{cc g}^{-1} \text{deg}^{-1}}$	$\dfrac{d_a}{\text{g cm}^{-3}}$	$\dfrac{d_c}{\text{g cm}^{-3}}$	$\dfrac{T_g}{°C}$	$\dfrac{T_d}{°C}$	$\dfrac{T_m}{°C}$	References
A 57					17_{1000}		450
					-110_{1000}		450
A 58	1.4/4.2			53			460
A 59	1.4/4.7			44			460
A 60				−23			93
A 61	1.5/4.3			11			460, 93
A 62				−50			380
A 63				−58			380
A 64				23			93
A 65				−24			380
A 66				−59			380
A 67	1.4/4.6			94			460
A 68				−58			380
A 69				19	-40_{200}		318, 256, (235)
A 70	1.1/3.7			60			460
A 71	1.6/5.0			47			460
A 72				−15b			327
A 73	2.1/6.8			22			460, (482)
A 74	1.7/4.9			24			460
A 75	1.8/4.7			30			460
A 76	1.8/4.4			37		151	460
A 77				−50			52
A 78				−55			52, 373
A 79	2.8/6.1	1.12		−21	$-5_{1.6}$		234,311,2,97,93
					$-50_{9.5}$		16,339,328,(482)
					-100_{11}		(318),(113),(235)
A 80				−50b			327
A 81				−55b			327
A 82				−20b			327
A 83	3.4/6.1			−6			460, (327)
A 84				−71			373

No.	Polymer	$\dfrac{\delta}{\mathrm{cal}^{1/2}\,\mathrm{cm}^{-3/2}}$	$\dfrac{\Delta H_u}{\mathrm{cal\ mol}^{-1}}$	n
A 85	**Acrylic acid,** 3-(ethylthio)propyl ester			
A 86	−, 2-(2, 2, 3, 3, 4, 4, 4-heptafluorobutoxy)ethyl ester			1.390_D^{25}
A 87	−, 2,2,3,3,4,4,4-heptafluorobutyl ester	6.7		1.367_D^{25}
A 88	−, heptyl ester			
A 89	−, hexadecyl ester			
A 90	−, 2,2,3,4,4,4-hexafluorobutyl ester			1.392_D^{25}
A 91	−, hexyl ester			
A 92	−, isobornyl ester	8.2		
A 93	−, isobutyl ester	9.0		
A 94	−, −, (isotactic)			
A 95	−, isopentyl ester			
A 96	−, isopropyl ester			
A 97	−, −, (isotactic)		1400	
A 98	−, −, (syndiotactic)			
A 99	−, m-(methoxycarbonyl)phenyl ester			
A 100	−, o-(methoxycarbonyl)phenyl ester			
A 101	−, p-(methoxycarbonyl)phenyl ester			
A 102	−, 2-methoxyethyl ester			1.463_D^{25}
A 103	−, p-methoxyphenyl ester			
A 104	−, 3-methoxypropyl ester			1.471_D^{25}
A 105	−, methyl ester	9.7		1.4793_D^{25}
		10.1		
		10.15		
A 106	−, 2-methylbutyl ester			
A 107	−, 1-methyl-4-ethyloctyl ester			
A 108	−, 1-methylheptyl ester			
A 109	−, 1-methylhexyl ester			
A 110	−, 2-methylpentyl ester			
A 111	−, 4-(methylthio)butyl ester			
A 112	−, 2-(methylthio)ethyl ester			

No.	$\dfrac{10^4\,d\bar{v}/dT}{\text{cc g}^{-1}\,\text{deg}^{-1}}$	$\dfrac{d_a}{\text{g cm}^{-3}}$	$\dfrac{d_c}{\text{g cm}^{-3}}$	$\dfrac{T_g}{°C}$	$\dfrac{T_d}{°C}$	$\dfrac{T_m}{°C}$	References
A 85				−76			373
A 86				−45			52
A 87				−30			52
A 88				−60			15
A 89				35			62, 15
A 90				−22			52
A 91				−60b			117
A 92				94			234, 318
A 93				−24b			234, 327
A 94		(1.05)	1.24			81	217
A 95				−45b			327
A 96	2.4/6.6			−3			460, 318, (235), (327)
	2.8/6.9						460
A 97				−11	180		523, 125, 318, 76
A 98				11	116		318, 522
A 99	1.3/4.5			38			460
A 100	1.5/4.5			46			460
A 101	1.8/3.7			67	>190		460
A 102				−50			52
A 103	1.7/4.7			48			460
A 104				−58			52, 373
A 105	2.7/5.6	1.22		9	$25_{1.2}$		234, 311, 2, 97, 16
	1.8/4.6				$-80_{9.5}$		339, 191, 16
							328, (318), (235), (450)
A 106				−32b			327
A 107				−40b			327
A 108				−45b			327
A 109				−38b			327
A 110				−38b			327
A 111				−70			373
A 112				−60			373

No.	Polymer	δ $\mathrm{cal}^{1/2}\,\mathrm{cm}^{-3/2}$	ΔH_u $\mathrm{cal\ mol}^{-1}$	n
A 113	**Acrylic acid,** 3-(methylthio)propyl ester			
A 114	—, 2,2,3,3,4,4,5,5,6,6,7,7,8,8,9,9,10,10,10-nonadecafluorodecyl ester			
A 115	—, 2,2,3,3,4,4,5,5,5-nonafluoropentyl ester			1.360_D^{25}
A 116	—, 2,2,3,3,4,4,5,5-octafluoropentyl ester			1.380_D^{25}
A 117	—, octyl ester			
A 118	—, pentachlorophenyl ester			
A 119	—, 2,2,3,3,4,4,5,5,6,6,7,7,8,8,8-pentadecafluoroöctyl ester			1.339_D^{25}
A 120	—, 2,2,3,3,3-pentafluoropropyl ester			1.385_D^{25}
A 121	—, phenyl ester			
A 122	—, 2-phenyethyl ester			
A 123	—, propyl ester	9.00		
A 124	—, tetradecyl ester			
A 125	—, 2-[2-(1,1,2,2-tetrafluoroethoxy)ethoxy]ethyl ester			1.422_D^{25}
A 126	—, 2-(1,1,2,2-tetrafluoroethoxy)ethyl ester			1.412_D^{25}
A 127	—, 2,2,3,3-tetrafluoro-3-(heptafluoropropoxy)propyl ester			1.346_D^{25}
A 128	—, 2,2,3,3-tetrafluoro-3-(nonafluorobutoxy)propyl ester			1.346_D^{25}
A 129	—, 2,2,3,3-tetrafluoro-3-(pentafluoroethoxy)propyl ester			1.348_D^{25}
A 130	—, 2,2,3,3-tetrafluoro-3-(trifluoromethoxy)propyl ester			1.360_D^{25}
A 131	—, 2-(2,2,2-trifluoroethoxy)ethyl ester			1.419_D^{25}
A 132	—, 2,2,2-trifluoroethyl ester			1.407_D^{25}
A 133	—, 2,2,3,3,4,4,5,5,6,6,6-undecafluorohexyl ester	6.5		1.356_D^{25}
A 134	—, **2-bromo-**, *sec*-butyl ester			1.542^{25}
A 135	—, —, cyclohexyl ester			1.547^{25}
A 136	—, —, methyl ester			1.5672_D^{25}
A 137	—, —, phenyl ester			1.612^{25}
A 138	—, **2-chloro-**, butyl ester			
A 139	—, —, *sec*-butyl ester			1.500^{25}
A 140	—, —, 2-chloroethyl ester			1.533^{25}
A 141	—, —, cyclohexyl ester			1.532^{25}
A 142	—, —, ethyl ester			1.502^{25}

No.	$10^4\ d\bar{v}/dT$ $cc\ g^{-1}\ deg^{-1}$	d_a $g\ cm^{-3}$	d_c $g\ cm^{-3}$	T_g °C	T_d °C	T_m °C	References
A 113				−65			373
A 114						100	52
A 115				−37			52
A 116				−35			52
A 117				−65s			33
A 118	0.94/1.9 0.85/3.0			147			460
A 119				−17		35	52
A 120				−26			52
A 121	1.6/5.1			55			460
A 122	1.6/5.2			−3			460
A 123				−44	14_{1000} -142_{1000}		339, 62, 450 450
A 124				20			62
A 125				−40			52
A 126				−22			52
A 127				−68			52
A 128				−68			52
A 129				−49			52
A 130				−55			52
A 131				−38			52
A 132				−10			52
A 133				−39			52
A 134							199
A 135							199
A 136							311, (199)
A 137							199
A 138		1.24		57s			175
A 139		1.24		74s			199, 175
A 140							199
A 141		1.25		114s	-30_{100}		199, 175
A 142		1.39		93s			199, 175

No.	Polymer	δ $\mathrm{cal}^{1/2}\,\mathrm{cm}^{-3/2}$	ΔH_u $\mathrm{cal\ mol}^{-1}$	n
A 143	**Acrylic acid, 2-chloro-**, isopropyl ester			
A 144	−, −, methyl ester	10.1		1.5172_D^{20}
A 145	−, −, propyl ester			
A 146	−, **2-cyano-**, butyl ester			
A 147	−, −, methyl ester	14.5		
		14.0		
A 148	−, **2-ethoxy-**, cyclohexyl ester			1.4969_D^{20}
A 149	−, **2-ethyl-**, ethyl ester			
A 150	−, −, methyl ester			
A 151	−, **2-fluoro-**, methyl ester			
A 152	−, **2-(fluoromethyl)-**, ethyl ester			
A 153	−, −, methyl ester			
A 154	−, **2-hexyl-**, methyl ester			
A 155	−, **2-isobutyl-**, methyl ester			
A 156	−, **2-methoxy-**, methyl ester			
A 157	−, **2-methyl-**. See **Methacrylic acid.**			
A 158	−, **3-methyl-**. See **Crotonic acid.**			
A 159	−, **2-propyl-**, methyl ester			
A 160	**Acrylonitrile**	15.4	1250	1.51_D^{20}
A 161	−, **2-ethyl-**			
A 162	−, **2-hexyl-**			
A 163	−, **2-isobutyl-**			
A 164	−, **2-methyl-**. See **Methacrylonitrile.**			
A 165	−, **2-propyl-**			
A 166	**Acrylophenone**			1.586_D^{20}
A 167	**Adipic acid,** polyamide with (4-aminobutyl)(3-aminopropyl)methylphosphine oxide			
A 168	−, polyamide with bis(3-aminopropyl)methylphosphine oxide			
A 169	−, polyamide with bis(3-aminopropyl)phenylphosphine oxide			

* Extrapolated value.

No.	$10^4\,d\bar{v}/dT$ $cc\ g^{-1}\ deg^{-1}$	d_a $g\ cm^{-3}$	d_c $g\ cm^{-3}$	T_g °C	T_d °C	T_m °C	References
A 143		1.27		90s			175
A 144		1.49		143	90_{200}		7, 311, 175, 93
		1.45					199
A 145		1.30		71s			175
A 146	1.1 / 5.1			85			430
A 147		1.304^{25}			170s		84
		1.289^{60}					234
A 148							311
A 149	1.5 / 5.7			27			460, 482
A 150				60s			117
A 151				131			93, (482)
A 152				54s	63_{60}		482
A 153				84	108_{60}		93, 482
A 154				12s			117
A 155				65s			117
A 156				60s			199
A 157							
A 158							
A 159				57s			117
A 160	1.39 / 2.95	1.178^0		85	105_1	317*	2, 26, 326, 32, 204, 16
		1.184^{20}					177, (18), (235)
A 161				110s			117
A 162				15s			117
A 163				55s			117
A 164							
A 165				60s			117
A 166				85s			311, 199
A 167		1.15		59			153
A 168		1.2		33			153
A 169				55			153

No.	Polymer	$\dfrac{\delta}{\mathrm{cal^{1/2}\ cm^{-3/2}}}$	$\dfrac{\varDelta H_u}{\mathrm{cal\ mol^{-1}}}$	n
A 170	**Adipic acid,** polyamide with 1,4-butanediamine			
A 171	–, polyamide with *cis*-1,4-cyclohexanebis(methylamine)			
A 172	–, polyamide with 1,10-decanediamine			
A 173	–, polyamide with 3,3'-diamino-N-methyldipropylamine			
A 174	–, polyamide with N,N'-dimethyl-1,6-hexanediamine			
A 175	–, polyamide with 1,12-dodecanediamine			
A 176	–, polyamide with 2,2'-(ethylenedioxy)bis(ethylamine)			
A 177	–, polyamide with fluorene-9,9-bis(propylamine)			
A 178	–, polyamide with 1,7-heptanediamine			
A 179	–, polyamide with 1,6-hexanediamine	13.6	4600	1.475 (α)
			8650	1.565 (β)
			10200	1.580 (γ)
			10590	
A 180	–, polyamide with 3,3'-(isobutylphosphinidene)bis(propylamine)			
A 181	–, polyamide with methanediamine			
A 182	–, polyamide with 3-methyl-1,6-hexanediamine			
A 183	–, polyamide with N-methyl-1,6-hexanediamine			
A 184	–, polyamide with 1,9-nonanediamine			
A 185	–, polyamide with 1,8-octanediamine			
A 186	–, polyamide with 3,3'-(octylphosphinidene)bis(propylamine)			
A 187	–, polyamide with 3,3'-oxybis(propylamine)			
A 188	–, polyamide with 1,5-pentanediamine			
A 189	–, polyamide with 3,3'-(phenylphosphinidene)bis(propylamine)			
A 190	–, polyamide with piperazine			
A 191	–, polyamide with spiro[3.3]heptane-2,6-diamine			
A 192	–, polyamide with 1,11-undecanediamine			
A 193	–, polyamide with m-xylene-α,α'-diamine			
A 194	–, polyanhydride			
A 195	–, polyester with p,p'-biphenol			
A 196	–, polyester with 2,2-bis(bromomethyl)-1,3-propanediol			

No.	$\dfrac{10^4 \, d\bar{v}/dT}{\text{cc g}^{-1}\,\text{deg}^{-1}}$	$\dfrac{d_a}{\text{g cm}^{-3}}$	$\dfrac{d_c}{\text{g cm}^{-3}}$	$\dfrac{T_g}{°C}$	$\dfrac{T_d}{°C}$	$\dfrac{T_m}{°C}$	References
A 170					43_{100}	295	564, 157, (122), (352)
A 171						246	410
A 172				40		240	122, 75, (157), (352)
A 173		1.12		5			153
A 174				−75			131, 157
A 175						230	75, (352)
A 176						160	157
A 177				120s			432
A 178			(1.104)	50	49	250	153, 122, 75, 348, (252)
A 179		1.09^{25}	1.26α	45	65_{89}	267	2, 55, 38, 563, 122, 20, 23
		1.069	1.24β		-41_{206}	165α	424, 162, 308, (69)
					-110_{187}		(352), (348)
							570
A 180				71			153
A 181						306	354
A 182						180	131, 157
A 183						145	131, 157
A 184						242	413, (75), (352)
A 185				45		254	122, 75, (348), (352)
A 186			(1.08)	12		135	153
A 187					34	213	252, (157)
A 188				45		258	122, 413, (352)
A 189		1.16		49			153
A 190						355	420, (85)
A 191						345	509
A 192						218	75
A 193		(1.22)	1.20	73s		244	360, 432, 75
			1.251				545
A 194						98s	155
A 195						300	371
A 196						120	462

No.	Polymer	$\dfrac{\delta}{\mathrm{cal}^{1/2}\,\mathrm{cm}^{-3/2}}$	$\dfrac{\Delta H_u}{\mathrm{cal\ mol}^{-1}}$	n
A 197	**Adipic acid,** polyester with 2,2-bis(chloromethyl)-1,3-propanediol			
A 198	—, polyester with 2-(bromomethyl)-2-(chloromethyl)-1,3-propanediol			
A 199	—, polyester with 1,4-butanediol			
A 200	—, polyester with *cis*-1,4-cyclohexanedimethanol			
A 201	—, polyester with *trans*-1,4-cyclohexanedimethanol			
A 202	—, polyester with 1,10-decanediol		3800	
			10200	
A 203	—, polyester with diethylene glycol			
A 204	—, polyester with 2,2-dimethyl-1,3-propanediol			
A 205	—, polyester with ethylene glycol		3800	
A 206	—, polyester with 1,6-hexanediol			
A 207	—, polyester with 4,4′-isopropylidenedi-*o*-cresol			
A 208	—, polyester with 4,4′-isopropylidenediphenol			
A 209	—, polyester with 4,4′-isopropylidenedi-2,6-xylenol			
A 210	—, polyester with 4,4′-methylenedi-*o*-cresol			
A 211	—, polyester with 4,4′-methylenediphenol			
A 212	—, polyester with 1,5-pentanediol			
A 213	—, polyester with 1,3-propanediol			
A 214	—, polyester with 3,3,3′,3′-tetramethyl-1,1′-spirobi[indan]-6,6′-diol			
A 215	—, polyester with 4,4′-(thiazolo[5,4-d]thiazole-2,5-diyl)bis(2-methoxyphenol)			
A 216	—, polyester with *p*-xylene-α,α'-diol			
A 217	**Allene,** 1,2-polymer			
A 218	—, **tetrafluoro**-, 1,2-polymer			
A 219	**Anisaldehyde,** polymercaptal with 1,6-hexanedithiol			
A 220	*m*-**Anisic acid,** α-**carboxy**-, polyanhydride			
A 221	—, **4,4′-(ethylenedioxy)di**-, polyester with ethylene glycol			
A 222	—, **4,4′-[oxybis(ethyleneoxy)]di**-, polyester with ethylene glycol			
A 223	—, **4,4′-(tetramethylenedioxy)di**-, polyester with ethylene glycol			
A 224	—, **4,4′-(trimethylenedioxy)di**-, polyester with ethylene glycol			
A 225	*p*-**Anisic acid,** *o*-**bromo**-α-**carboxy**-, polyanhydride			

No.	$10^4 \, d\bar{v}/dT$ cc g^{-1} deg^{-1}	d_a g cm^{-3}	d_c g cm^{-3}	T_g °C	T_d °C	T_m °C	References
A 197						108	462
A 198						111	462
A 199		(1.0188)[20]		−68		54	330, 51, (343)
A 200						55	384
A 201						124	384
A 202	/7.34	0.9732[109]		−56		82	428, 243, 51
							37, 307, (55)
A 203	/6.36	1.125[109]		−46			243, 51, (145)
A 204						37	358
A 205		(1.1825)[20]	1.274	−50		55	155, 330, 195, 51, 102
			(1.28)				411, (131)
A 206			(1.0206)[20]			57	330, (55)
A 207				33.5s			371
A 208				68			115, (371)
A 209				106			115
A 210				30			115
A 211				60.5s		140	371, (115)
A 212				−69			51
A 213				−59		46	51, 358, (195)
A 214					140$_{0.1}$		455
A 215						293	361
A 216						81	384
A 217		(0.97)[25]	1.07			122	456, 441, (436)
A 218		(1.99)	2.02			126	81
A 219						130	241
A 220						134	229
A 221						210	253
A 222						118	253
A 223						117	253
A 224						105	253
A 225						92	229

No.	Polymer	$\dfrac{\delta}{cal^{1/2}\ cm^{-3/2}}$	$\dfrac{\Delta H_u}{cal\ mol^{-1}}$	n
A 226	*p*-Anisic acid, α-carboxy-, polyester with ethylene glycol			
A 227	−, −, polyester with 1,6-hexanediol			
A 228	Anisole, 3-methyl-4-vinyl-. See Styrene, 4-methoxy-2-methyl-.			
A 229	−, vinyl-. See Styrene, methoxy-.			
A 230	Azelaic acid, polyamide with 1,4-butanediamine			
A 231	−, polyamide with *cis*-1,4-cyclohexanebis(methylamine)			
A 232	−, polyamide with *trans*-1,4-cyclohexanebis(methylamine)			
A 233	−, polyamide with 1,10-decanediamine		8700	
			8800	
A 234	−, polyamide with 1,7-heptanediamine			
A 235	−, polyamide with 1,6-hexanediamine			
A 236	−, polyamide with 1,9-nonanediamine			
A 237	−, polyamide with 1,8-octanediamine			
A 238	−, polyamide with 1,5-pentanediamine			
A 239	−, polyamide with 2,2'-*p*-phenylenebis(ethylamine)			
A 240	−, polyamide with piperazine			
A 241	−, polyamide with spiro[3.3]heptane-2,6-diamine			
A 242	−, polyamide with *p*-xylene-α,α'-diamine			
A 243	−, polyester with 1,4-butanediol			
A 244	−, polyester with *cis*-1,4-cyclohexanedimethanol			
A 245	−, polyester with *trans*-1,4-cyclohexanedimethanol			
A 246	−, polyester with 1,10-decanediol		10000	
			10100	
A 247	−, polyester with diethylene glycol			
A 248	−, polyester with 2,2-dimethyl-1,3-propanediol			
A 249	−, polyester with ethylene glycol			
A 250	−, polyester with 1,6-hexanediol			
A 251	−, polyester with 1,9-nonanediol		10300	
A 252	−, polyester with 1,5-pentanediol			
A 253	−, polyester with 1,3-propanediol			
A 254	−, polyester with *p*-xylene-α,α'-diol			

No.	$10^4 \, d\bar{v}/dT$ cc g^{-1} deg^{-1}	d_a g cm^{-3}	d_c g cm^{-3}	T_g °C	T_d °C	T_m °C	References
A 226						140	55
A 227						50	55
A 228							
A 229							
A 230			$(1.097)^{25}$		-41_{200}	223	162, 352
A 231						195	410
A 232						275	410
A 233	/6.6	0.9074^{218}	$(1.044)^{25}$		-36_{198}	214	130, 307, 162, 307
							307
A 234			$(1.060)^{25}$		-38_{203}	201s	162, 348
A 235			$(1.076)^{25}$		-40_{199}	226s	162, 348, (352)
A 236			$(1.042)^{25}$		-38_{92}	177	162, 157, (195), (307), (352)
A 237			$(1.058)^{25}$		-37_{207}	206s	162, 348
A 238						179	157, (352)
A 239			$(1.140)^{25}$			290	529
A 240						148s	85
A 241						225	509
A 242			$(1.152)^{25}$			282	529, (410)
A 243						39	343
A 244						41	384
A 245						50	384
A 246		0.938^{100}				69	130, 307
							23
A 247				-68		0	51, 343
A 248						0	358
A 249			1.220	-45		46	195, 102
A 250			$(1.0312)^{20}$			53	330
A 251						65	307, 23
A 252						41	343
A 253						60	358, (343), (195)
A 254						79	384

No.	Polymer	$\dfrac{\delta}{\mathrm{cal}^{1/2}\,\mathrm{cm}^{-3/2}}$	$\dfrac{\varDelta H_u}{\mathrm{cal\ mol}^{-1}}$	n
B 1	**Behenic acid.** See **Docosanoic acid.**			
B 2	**Benzaldehyde,** copolymer *(alt)* with dimethylketene. See **Hydracrylic acid, 2,2-dimethyl-3-phenyl-**, polyester.			
B 3	**Benzaldehyde,** polymercaptal with 1,10-decanedithiol			
B 4	−, polymercaptal with 1,6-hexanedithiol			
B 5	−, *p*-**bromo**-, polymercaptal with 1,6-hexanedithiol			
B 6	−, *m*-**nitro**-, polymercaptal with 1,10-decanedithiol			
B 7	**Benzene,** 1,3-ethylene polymer *(alt)*			
B 8	−, 1,4-ethylene polymer *(alt)*			
B 9	−, **allyl**-, (stereoregular)			
B 10	−, **1,4-bis(hydroxymethyl)**-. See *p*-**Xylene-**α,α'-**diol.**			
B 11	−, **1,4-dihydroxy**-. See **Hydroquinone.**			
B 12	−, **epithioethyl**-, poly(thioether)	9.3		
B 13	−, **epoxyethyl**-, polyether			
B 14	−, **isopropenyl**-. See **Styrene,** α-**methyl**-.			
B 15	−, **vinyl**-. See **Styrene.**			
B 16	*p*-**Benzenediacetic acid,** polyamide with 1,10-decanediamine			
B 17	−, polyamide with 1,12-dodecanediamine			
B 18	−, polyamide with 1,6-hexanediamine			
B 19	−, polyamide with 1,18-octadecanediamine			
B 20	−, polyamide with 1,8-octanediamine			
B 21	−, polyamide with 1,14-tetradecanediamine			
B 22	−, polyanhydride			
B 23	−, polyester with *cis*-1,4-cyclohexanedimethanol			
B 24	−, polyester with ethylene glycol			
B 25	−, polyester with 1,3-propanediol			
B 26	−, polyester with *p*-xylene-α,α'-diol			
B 27	**1,2-Benzenedicarboxylic acid.** See **Phthalic acid.**			
B 28	**1,3-Benzenedicarboxylic acid.** See **Isophthalic acid.**			

No.	$\dfrac{10^4 \, d\bar{v}/dT}{cc \, g^{-1} \, deg^{-1}}$	d_a g cm^{-3}	d_c g cm^{-3}	T_g °C	T_d °C	T_m °C	References
B 1							
B 2							
B 3						135	241
B 4						115	241
B 5						72	241
B 6						75	241
B 7						80	465
B 8	(1.148)	1.084α				440β	232, 474, 465, (247)
		1.248β				220α	
B 9	1.046	(1.051)		60	$70_{0.1}$	240	160, 147, 113, 76, (117), (292)
B 10							
B 11							
B 12				60s			181
B 13	1.15			25s		149	478, 78
B 14							
B 15							
B 16			$(1.130)^{25}$			265	529, (157)
B 17			$(1.095)^{25}$			256	529
B 18			$(1.182)^{25}$			300	529
B 19			$(1.036)^{25}$		78_{110}	225	529
B 20			$(1.152)^{25}$		110_{110}	280	529
B 21			$(1.075)^{25}$			235	529
B 22						92	229
B 23						55	384
B 24						137s	155
B 25						58	384
B 26						146	57, 157
B 27							
B 28							

No.	Polymer	$\dfrac{\delta}{\text{cal}^{1/2}\ \text{cm}^{-3/2}}$	$\dfrac{\Delta H_u}{\text{cal mol}^{-1}}$	n
B 29	**1,4-Benzenedicarboxylic acid.** See **Terephthalic acid.**			
B 30	**1,4-Benzenediol.** See **Hydroquinone.**			
B 31	*m*-**Benzenedisulfonic acid,** polyester with 4,4′-isopropylidenediphenol			
B 32	*p*-**Benzenedisulfonic acid,** polyester with 4,4′-isopropylidenediphenol			
B 33	**Benzenesulfonic acid, carboxy-.** See **Benzoic acid, sulfo-.**			
B 34	−, **4,4′-oxydi-,** polyester with 4,4′-*sec*-butylidenediphenol			
B 35	−, −, polyester with 4,4′- (1-ethylpropylidene)diphenol			
B 36	−, −, polyester with 4,4′-isopropylidenediphenol			
B 37	**1,2,4,5-Benzenetetracarboxylic acid.** See **Pyromellitic acid.**			
B 38	**Benzoic acid,** vinyl ester			1.5775_D^{20}
B 39	−, *p*-(**aminomethyl**)-, polyamide			
B 40	−, **4,4′-*sec*-butylidenedi-,** polyamide with 1,4-butanediamine			
B 41	−, −, polyamide with 1,6-hexanediamine			
B 42	−, −, polyamide with piperazine			
B 43	−, −, polyester with ethylene glycol			
B 44	−, −, polyester with 4,4′-isopropylidenediphenol			
B 45	−, −, polyester with 1,3-propanediol			
B 46	−, **4,4′-(*sec*-butylidenedioxy)di-,** polyester with ethylene glycol			
B 47	−, **4,4′-carbonylidi-,** polyester with 4,4′-isopropylidenediphenol			
B 48	−, *p*-(**4-carboxybutoxy**)-, polyester with ethylene glycol			
B 49	−, −, polyester with 1,6-hexanediol			
B 50	−, **carboxymethoxy-.** See **Anisic acid, α-carboxy-.**			
B 51	−, *p*-(**5-carboxypentyloxy**)-, polyester with ethylene glycol			
B 52	−, −, polyester with 1,6-hexanediol			
B 53	−, *p*-(**3-carboxypropoxy**)-, polyester with ethylene glycol			
B 54	−, **4,4′-(dichloromethylene)di -,** polyester with 4,4′-isopropylidenediphenol			
B 55	−, **4,4′-ethylenedi-,** polyanhydride			
B 56	−, −, polyester with ethylene glycol			
B 57	−, **4,4′-(ethylenediamino)di-,** polyester with ethylene glycol			
B 58	−, **3,3′-(ethylenedioxy)di-,** polyanhydride			
B 59	−, −, polyester with ethylene glycol			

No.	$\dfrac{10^4 \, d\bar{v}/dT}{cc \ g^{-1} \ deg^{-1}}$	$\dfrac{d_a}{g \ cm^{-3}}$	$\dfrac{d_c}{g \ cm^{-3}}$	$\dfrac{T_g}{°C}$	$\dfrac{T_d}{°C}$	$\dfrac{T_m}{°C}$	References
B 29							
B 30							
B 31				115		115	
B 32				127		115	
B 33							
B 34				124		115	
B 35				132		115	
B 36				121		115	
B 37							
B 38				68		311, 93	
B 39				300		210	
B 40				182		99	
B 41				164		99	
B 42				232		99	
B 43				109		99	
B 44				211		99	
B 45				70		99	
B 46	$(1.72)/(3.24) \, (1.297)^0$			53		32	
B 47					240s	154	
B 48					55	55	
B 49					60	55	
B 50							
B 51					45	55	
B 52					60	55	
B 53					85	55	
B 54				260s		154	
B 55					> 340	379	
B 56					220	180, (258)	
B 57					273	180	
B 58					237	379	
B 59					141	251	

No.	Polymer	$\dfrac{\delta}{\text{cal}^{1/2}\,\text{cm}^{-3/2}}$	$\dfrac{\Delta H_u}{\text{cal mol}^{-1}}$	n
B 60	**Benzoic acid, 4,4'-(ethylenedioxy)di-**, polyanhydride			
B 61	−, −, polyester with 1,4-butanediol			
B 62	−, −, polyester with 1,10-decanediol			
B 63	−, −, polyester with ethylene glycol			
B 64	−, −, polyester with 1,6-hexanediol			
B 65	−, −, polyester with 1,18-octadecanediol			
B 66	−, −, polyester with 1,5-pentanediol			
B 67	−, −, polyester with 1,3-propanediol			
B 68	−, **4,4'-(ethylenedithio)di-**, polyester with ethylene glycol			
B 69	−, **4,4'-hexamethylenedi-**, polyanhydride			
B 70	−, **3,3'-(hexamethylenedioxy)di-**, polyanhydride			
B 71	−, **4,4'-(hexamethylenedioxy)di-**, polyanhydride			
B 72	−, −, polyester with ethylene glycol			
B 73	−, *p*-**hydroxy-**, polyester			
B 74	−, *p*-**(2-hydroxyethoxy)-**, polyester			
B 75	−, *p*-**(3-hydroxypropoxy)-**, polyester			
B 76	−, **4,4'-isopropylidenedi-**, polyamide with 1,6-hexanediamine			
B 77	−, −, polyanhydride			
B 78	−, −, polyester with 4,4'-carbonyldiphenol			
B 79	−, −, polyester with 4,4'-isopropylidenediphenol			
B 80	−, −, polyester with 4,4'-methylenediphenol			
B 81	−, **methoxy-**. See **Anisic acid.**			
B 82	−, **methyl-**. See **Toluic acid.**			
B 83	−, **4,4'-methylenedi-**, polyanhydride			
B 84	−, −, polyester with ethylene glycol			
B 85	−, −, polyester with 4,4'-(1-ethylpropylidene)diphenol			
B 86	−, −, polyester with 4,4'-isopropylidenediphenol			
B 87	−, **4,4'-(methylenedioxy)di-**, polyanhydride			
B 88	−, **4,4'-[oxybis(ethyleneoxy)]di-**, polyanhydride			
B 89	−, **4,4'-[oxybis(methyleneoxy)]di-**, polyanhydride			
B 90	−, **3,3'-oxydi-**, polyester with 4,4'-isopropylidenediphenol			

No.	$\dfrac{10^4\ d\bar{v}/dT}{cc\ g^{-1}\ deg^{-1}}$	$\dfrac{d_a}{g\ cm^{-3}}$	$\dfrac{d_c}{g\ cm^{-3}}$	$\dfrac{T_g}{°C}$	$\dfrac{T_d}{°C}$	$\dfrac{T_m}{°C}$	References
B 60				45		215	379, 229
B 61						180	55, 180
B 62						135	55
B 63						240	55, 157, 251
B 64						175	55
B 65						122	55
B 66						100	180
B 67						160	180
B 68						200	180, (157)
B 69				<20		151	379
B 70						157	379
B 71		$(1.233)^{20}$				157	379
B 72						170	180
B 73						>350	180
B 74						203	400, (180), (155)
B 75						211	400, (155)
B 76				140			99
B 77				60		240	379, 229
B 78				240s			154
B 79				230s			154
B 80				140s			154
B 81							
B 82							
B 83				122	332		379
B 84				136	220		157
B 85				200s			154
B 86				205s			154
B 87				84	220		379
B 88				41	190		379
B 89				52	192		379
B 90				85s			154

No.	Polymer	$\dfrac{\delta}{\text{cal}^{1/2}\,\text{cm}^{-3/2}}$	$\dfrac{\Delta H_u}{\text{cal mol}^{-1}}$	n
B 91	**Benzoic acid, 4,4′-oxydi-**, polyanhydride			
B 92	—, **4,4′-oxydi-**, polyester with 4,4′-benzylidenediphenol			
B 93	—, —, polyester with 4,4′-*sec*-butylidenediphenol			
B 94	—, —, polyester with 4,4′-ethylidenediphenol			
B 95	—, —, polyester with 4,4′-(1-ethylpropylidene)diphenol			
B 96	—, —, polyester with 4,4′-isopropylidenediphenol			
B 97	—, —, polyester with 4,4′-(α-methylbenzylidene)diphenol			
B 98	—, **4,4′-pentamethylenedi-**, polyanhydride			
B 99	—, **3,3′-(pentamethylenedioxy)di-**, polyanhydride			
B 100	—, **4,4′-(pentamethylenedioxy)di-**, polyanhydride			
B 101	—, —, polyester with ethylene glycol			
B 102	—, *m*-**sulfo-**, polyester with 4,4′-isopropylidenediphenol			
B 103	—, *p*-**sulfo-**, polyester with 4,4′-isopropylidenediphenol			
B 104	—, **4,4′-sulfonyldi-**, polyester with 2,2′-(2,5-di-*tert*-butyl-*p*-phenylenedioxy)diethanol			
B 105	—, —, polyester with 4,4′-isopropylidenediphenol			
B 106	—, **4,4′-tetramethylenedi-**, polyanhydride			
B 107	—, —, polyester with ethylene glycol			
B 108	—, **3,3′-(tetramethylenedioxy)di-**, polyanhydride			
B 109	—, —, polyester with ethylene glycol			
B 110	—, **4,4′-(tetramethylenedioxy)di-**, polyanhydride			
B 111	—, **4,4′-(tetramethylenedioxy)di-**, polyester with ethylene glycol			
B 112	—, **4,4′-(tetramethylenedithio)di-**, polyanhydride			
B 113	—, **4,4′-thiodi-**, polyester with ethylene glycol			
B 114	—, **4,4′-trimethylenedi-**, polyanhydride			
B 115	—, **3,3′-(trimethylenedioxy)di-**, polyanhydride			
B 116	—, **4,4′-(trimethylenedioxy)di-**, polyanhydride			
B 117	—, —, polyester with ethylene glycol			
B 118	—, *p*-**vinyl-**, methyl ester. See **Styrene, *p*-methoxycarbonyl-**.			
B 119	**2,2′-Bibenzoic acid.** See **Diphenic acid.**			
B 120	**Bibenzoic acid,** other isomers. See **Biphenyldicarboxylic acid.**			
B 121	**Bicyclo[2·2·1]hept-2-ene.** See **Cyclopentane,** 1,3-*trans*-vinylene polymer *(alt)*.			

No.	$\dfrac{10^4 \, d\bar{v}/dT}{\text{cc g}^{-1}\,\text{deg}^{-1}}$	$\dfrac{d_a}{\text{g cm}^{-3}}$	$\dfrac{d_c}{\text{g cm}^{-3}}$	$\dfrac{T_g}{^\circ\text{C}}$	$\dfrac{T_d}{^\circ\text{C}}$	$\dfrac{T_m}{^\circ\text{C}}$	References
B 91						296	379
B 92				200s			154
B 93				175s			154
B 94				130s			154
B 95				140s			154
B 96				220s			154
B 97				270s			154
B 98				39	118		379
B 99				61	176		379
B 100				53	188		379
B 101					150		180
B 102				133			115
B 103				135			115
B 104					240		236
B 105				220s			154
B 106				46	263		379
B 107					170		157
B 108				<20	199		379
B 109					100		251
B 110		$(1.266)^{20}$	$(1.301)^{20}$	75	204		379
B 111					252		155, (180)
B 112				62	206		379
B 113					200		157
B 114					215		379
B 115				53	197		379
B 116				95	267		379
B 117					190		180
B 118							
B 119							
B 120							
B 121							

No.	Polymer	$\dfrac{\delta}{\text{cal}^{1/2}\ \text{cm}^{-3/2}}$	$\dfrac{\Delta H_u}{\text{cal mol}^{-1}}$	n
B 122	**Bicyclo[2·2·1]hepta-2,5-diene-2-carboxylic acid.** See 2,5-**Norbornadiene-2-carboxylic acid.**			
B 123	**Biphenyl, 4-vinyl-**			
B 124	**2,2′-Biphenyldicarboxylic acid.** See **Diphenic acid.**			
B 125	**3,3′-Biphenyldicarboxylic acid,** polyester with ethylene glycol			
B 126	**4,4′-Biphenyldicarboxylic acid,** polyester with 1,4-butanediol			
B 127	−, polyester with 2,2′-(2,5-di-*tert*-butyl-*p*-phenylenedioxy)diethanol			
B 128	−, polyester with diethylene glycol			
B 129	−, polyester with ethylene glycol			
B 130	−, polyester with 1,6-hexanediol			
B 131	−, **2,2′-dimethyl-,** polyester with ethylene glycol			
B 132	**Brassylic acid.** See **Tridecanedioic acid.**			
B 133	**1,3-Butadiene**	8.1 8.38 8.5		1.5154_D^{25}
B 134	−, 1,2-polymer (isotactic)			
B 135	−, 1,2-polymer (syndiotactic)			
B 136	−, *cis*-1,4-polymer		2000 2200	
B 137	−, *trans*-1,4-polymer		2400 I 1100 II	
B 138	−, 2-*tert*-**butyl-**			$1.5060_D^{24.6}$
B 139	−, −, *cis*-1,4-polymer			
B 140	−, **2-chloro-.** See **Chloroprene.**			
B 141	−, **2-decyl-**			$1.4899_D^{20.5}$
B 142	−, **2,3-dimethyl-**			
B 143	−, −, *cis*-1,4-polymer			
B 144	−, −, *trans*-1,4-polymer			
B 145	−, **2-ethyl-**			

* Major (glass) transition

No.	$10^4\, d\bar{v}/dT$ cc g^{-1} deg^{-1}	d_a g cm^{-3}	d_c g cm^{-3}	T_g °C	T_d °C	T_m °C	References
B 122							
B 123				138s		96	
B 124							
B 125					156s	180	
B 126				>100	280	258	
B 127					270	236	
B 128				80	346	32, 34	
B 129				>100	355	258, 180, (157)	
B 130					230	258, (157)	
B 131				73		258	
B 132							
B 133	/7.8 /7.50	0.8920^{25}		−58			2, 6187, 112 7 9
B 134			0.96	−65	125		549, 76
B 135		(0.92)	0.963		154		60, 549
B 136			1.01	−102	6.3		444, 549, 469, 324, (531), 331, (112), (237), (280)
B 137			1.02 I 0.93^{80}II	−10 −48*	$-14_{10,000}$	148 I 109 II 80III 52IV	331, 322, 112, 550, 549 476, (222) 280, (223) 112, (237)
B 138				20			375, 399
B 139		(0.88)	0.906	25	106		129
B 140							
B 141				−53	−40		129
B 142				−11			1, (61)
B 143					198	282	
B 144					272		427, (136), (287)
B 145				−70			117

No.	Polymer	δ $\overline{cal^{1/2}\,cm^{-3/2}}$	ΔH_u $\overline{cal\,mol^{-1}}$	n
B 146	**1,3-Butadiene, 2-heptyl-**			1.5000_D^2
B 147	−, **2-hexyl-**			
B 148	−, **2-isopropyl-**			1.5028_D^{30}
B 149	−, **1-methoxy-**			
B 150	−, −, *trans*-1,4-polymer (stereoregular)			
B 151	−, **2-methyl-**. See **Isoprene.**			
B 152	−, **2-pentyl-**			
B 153	−, **2-(tributyltin)-**, *cis*-1,4-polymer			
B 154	**Butadiene monoxide.** See **1-Butene, 3,4-epoxy-.**			
B 155	**Butanal.** See **Butyraldehyde.**			
B 156	**Butane, 2,2-bis(*p*-carboxyphenoxy)-**. See **Benzoic acid, 4,4′-(*sec*-butylidenedioxy)di-.**			
B 157	−, **2,2-bis(*p*-carboxyphenyl)-**. See **Benzoic acid, 4,4′-*sec*-butylidenedi-.**			
B 158	−, **2,3-dimethyl-2,3-epoxy-**, polyether			
B 159	−, **2,3-epoxy-**, *cis*-, *threo*-polyether (diisotactic)			
B 160	−, −, *trans*-, *erythro*-polyether (diisotactic)			
B 161	**Butanedioic acid.** See **Succinic acid.**			
B 162	**Butanoic acid.** See **Butyric acid.**			
B 163	**1-Butene**			
B 164	−, (isotactic)		1675 I	
			1450 I	
			1500 II	
			1570 II	
			1550 III	
B 165	−, **4-cyclohexyl-**, (stereoregular)			
B 166	−, **3,3-dimethyl-**, (stereoregular)			
B 167	−, **3,4-epoxy-**, polyether			
B 168	−, **4-iodo-**			
B 169	−, **3-methyl-**, (isotactic)			

** Second-order transition

No.	$10^4\,d\bar{v}/dT$ $cc\ g^{-1}\ deg^{-1}$	d_a $g\ cm^{-3}$	d_c $g\ cm^{-3}$	T_g $°C$	T_d $°C$	T_m $°C$	References
B 146				−83			129
B 147				−83			117
B 148				−47			129
B 149				−17			469
B 150						118	246
B 151							
B 152				−55			117
B 153				−80			393
B 154							
B 155							
B 156							
B 157							
B 158					>300		143
B 159		(1.068)	1.168			162	506, 78
B 160		(1.015)	1.099			114	506, 78
B 161							
B 162							
B 163				−45			76
B 164	$/8.78$	0.8595^{25}	0.9507^{25} I	−24	$-16_{0.42}$	142 I	386, 287, 451, 237, 314
	$/8.07^{100}$	0.864^{30}	0.945^{30} I		-125_{430}	126 II	291, 294, 97, (60)
		0.87	0.886^{25} II			106 III	275, (46), (47)
		0.821^{100}	0.862^{100} II			65**IV	423, (388), (40)
							(408), (344), (141)
B 165				40s	$40_{0.1}$	170	292, 113, (401)
B 166				64s		300	292, 76
B 167						74	78
B 168					70_{240}		486
					-90_{380}		486
B 169		(0.90)	0.932	53s	$50_{0.1}$	310	76, 383, 292, 113, 117, 525
			0.92		-120_{1600}		438, 504, (95), (417)

No.	Polymer	$\dfrac{\delta}{\text{cal}^{1/2}\ \text{cm}^{-3/2}}$	$\dfrac{\Delta H_u}{\text{cal mol}^{-1}}$	n
B 170	1-**Butene, 3-phenyl-**, (stereoregular)			
B 171	—, **4-phenyl-**, (stereoregular)			
B 172	—, **4-*o*-tolyl-**, (stereoregular)			
B 173	—, **4-*p*-tolyl-**, (stereoregular)			
B 174	—, **4,4,4-trifluoro-**, (stereoregular)			
B 175	—, **3-trifluoromethyl-**, (stereoregular)			
B 176	2-**Butene,** (stereoregular)			
B 177	*cis*-2-**Butene,** copolymer *(alt)* with ethylene			
B 178	1-**Butene-2-carboxylic acid.** See **Acrylic acid, 2-ethyl-.**			
B 179	*trans*-**Butenedioic acid.** See **Fumaric acid.**			
B 180	**Butene oxide.** See **Butane, 2,3-epoxy-.**			
B 181	**3-Butene-2-one**			1.50_D
B 182	—, (isotactic)			
B 183	—, **3-fluoro-**			
B 184	—, **3-methyl-**			1.5200_D^{30}
B 185	—, —, (stereoregular)			
B 186	**Butyraldehyde,** polyacetal (isotactic)			
B 187	—, polymercaptal with 1.6-hexanedithiol			
B 188	**Butyric acid,** vinyl ester			
B 189	**Butyric acid, 4-amino-,** polyamide			
B 190	—, **3-(2-amino-1-methylethoxy)-,** polyamide			
B 191	—, **heptafluoro-,** vinyl ester			1.363_D^{25}
B 192	—, **3-hydroxy-,** D(-), polyester (isotactic)			
B 193	—, **4,4'oxydi-,** polyamide with 1,6-hexanediamine			
B 194	—, —, polyamide with 5,5'-oxybis(pentylamine)			
B 195	—, —, polyamide with *p*-xylene-α,α'-diamine			
B 196	—, **4,4'-thiodi-,** polyamide with 1,6-hexanediamine			

No.	$10^4 \, d\bar{v}/dT$ cc g⁻¹ deg⁻¹	d_a g cm⁻³	d_c g cm⁻³	T_g °C	T_d °C	T_m °C	References
B 170						>360	76, 117
B 171			(1.04)	10	$40_{0.1}$	168	147, 160, 113, (292), (117)
B 172						239	147
B 173						196	147
B 174						263	6
B 175						300	6
B 176			(0.92)			190	449, 132
B 177	0.87		0.95			135	134, 313
B 178							
B 179							
B 180							
B 181	1.12			40s			199
B 182	(1.17)		1.216			170	337, 409, (219)
B 183				143			422
B 184	1.15^{20}			80			311, 199, 271
B 185	1.12 ^{25}		$(1.15)^{25}$	114		240	271, (409)
B 186			0.997		20_1	225	104, 405
					-30_1		405
B 187						40	241
B 188					4_{15}		97
					-165_{1000}		450
B 189	$(1.25)^{26}$		1.37α		$-60_{0.30}$	265	544, 329, (162)
					$-146_{0.31}$		329
B 190						210	528
B 191					27		359
B 192	1.23		1.262			184	285, 515
B 193						187	527
B 194						138	527
B 195						241	527
B 196						200	157

No.	Polymer	$\dfrac{\delta}{\text{cal}^{1/2}\,\text{cm}^{-3/2}}$	$\dfrac{\varDelta H_u}{\text{cal mol}^{-1}}$	n
C 1	**Capric acid.** See **Decanoic acid.**			
C 2	**Caproic acid.** See **Hexanoic acid.**			
C 3	**Caprolactam.** See **Hexanoic acid, 6-amino-.**			
C 4	**Caprylic acid.** See **Octanoic acid.**			
C 5	**Carbamic acid, butyl-,** polyamide			
C 6	−, **(6-carboxyhexyl)-**, polyester with 1,4-butanediol			
C 7	−, −, polyester with 1,10-decanediol			
C 8	−, −, polyester with 1,6-hexanediol			
C 9	−, −, polyester with 1,5-pentanediol			
C 10	−, **hexamethylenebis[phenyl-**, polyester with 4,4′-isopropylidenebis (2,6-dichlorophenol)			
C 11	−, −, polyester with 4,4′-isopropylidenediphenol			
C 12	−, **(isopropylidenedi-p-phenylene)bis[phenyl-**, polyester with 4,4′-isopropylidenediphenol			
C 13	−, **(methylenedi-p-phenylene)bis[ethyl-**, polyester with 4,4′-isopropylidenebis(2,6-dichlorophenol)			
C 14	−, −, polyester with 4,4′-isopropylidenediphenol			
C 15	−, **(methylenedi-p-phenylene)bis[methyl-**, polyester with 4,4′-isopropylidenebis(2,6-dichlorophenol)			
C 16	−, −, polyester with 4,4′-isopropylidenediphenol			
C 17	−, −, polyester with 4,4′-sulfonyldiphenol			
C 18	−, **(methylenedi-p-phenylene)bis [phenyl-**, polyester with 4,4′-isopropylidenediphenol			
C 19	−, **p-phenylenebis[phenyl-**, polyester with 4,4′-isopropylidenediphenol			
C 20	**Carbanilic acid,** polyamide			
C 21	−, **m-carboxy-**, polyester with 1,4-butanediol			
C 22	−, −, polyester with 1,10-decanediol			
C 23	−, −, polyester with 1,6-hexanediol			
C 24	−, −, polyester with 1,5-pentanediol			
C 25	−, **p-carboxy-**, polyester vith 1,4-butanediol			
C 26	−, −, polyester with 1,10-decanediol			
C 27	−, −, polyester with 1,5-pentanediol			

No.	$\dfrac{10^4 \, d\bar{v}/dT}{cc \, g^{-1} \, deg^{-1}}$	$\dfrac{d_a}{g \, cm^{-3}}$	$\dfrac{d_c}{g \, cm^{-3}}$	$\dfrac{T_g}{°C}$	$\dfrac{T_d}{°C}$	$\dfrac{T_m}{°C}$	References
C 1							
C 2							
C 3							
C 4							
C 5						175	240
C 6						83	535
C 7						91	535
C 8						75	535
C 9						70	535
C 10					130		491
C 11					90		491
C 12					190		491
C 13					200		491
C 14					130		491
C 15					210		491
C 16					155		491
C 17					190		491
C 18					160		491
C 19					180		491
C 20						275	240
C 21						60	535
C 22						127	535
C 23						96	535
C 24						80	535
C 25						186	535
C 26						158	535
C 27						160	535

No.	Polymer	$\dfrac{\delta}{\mathrm{cal}^{1/2}\,\mathrm{cm}^{-3/2}}$	$\dfrac{\Delta H_u}{\mathrm{cal\,mol}^{-1}}$	n
C 28	**Carbanilic acid,** N,N'-**hexamethylenedi-**.			
	See **Carbamic acid, hexamethylenebis [phenyl-**.			
C 29	−, 4,4′-methylenebis [N-methyl-.			
	See **Carbamic acid, (methylenedi-p-phenylene)bis [methyl-**.			
C 30	−, p-phenylenedi-. See **Carbamic acid, p-phenylenebis [phenyl-**.			
C 31	**Carbazole, 9-vinyl-**			1.683_D^{20}
C 32	**Carbonic acid,** cyclic phenylvinylene ester			
C 33	−, polyester with 4,4′-benzylidenedi-m-cresol			
C 34	−, polyester with 4,4′-benzylidenediphenol			
C 35	−, polyester with 4,4′-(4,4′-biphenylylenedioxy)diphenol			
C 36	−, polyester with 4,4′-butylidenediphenol			1.5792
C 37	−, polyester with 4,4′-sec-butylidenediphenol			1.5827
C 38	−, polyester with 4,4′-cyclohexylidenebis(2-chlorophenol)			
C 39	−, polyester with 4,4′-cyclohexylidenebis(2,6-dichlorophenol)			1.5858
C 40	−, polyester with 4,4′-cyclohexylidenedi-o-cresol			
C 41	−, polyester with 4,4′-cyclohexylidenediphenol			1.5900
C 42	−, polyester with 4,4′-cyclopentylidenediphenol			1.5993
C 43	−, polyester with 4,4′-decamethylenediphenol			
C 44	−, polyester with α,α'-diethyl-4,4′-stilbenediol			
C 45	−, polyester with 4,4′-(1,3-dimethylbutylidene)diphenol			1.5671
C 46	−, polyester *(alt)* with 2,2-dimethyl-1,3-propanediol and 4,4′-(thiazolo[5,4-d]thiazole-2,5-diyl)bis(2-methoxyphenol)			
C 47	−, polyester *(alt)* with 2,2-dimethyl-1,3-propanediol and 3,3′-(thiazolo[5,4-d]thiazole-2,5-diyl)diphenol			
C 48	−, polyester *(alt)* with 2,2-dimethyl-1,3-propanediol and 4,4′-(thiazolo[5,4-d]thiazole-2,5-diyl)diphenol			
C 49	−, polyester with 4,4′-(diphenylmethylene)diphenol			1.6539
C 50	−, polyester with 4,4′-ethylenediphenol			
C 51	−, polyester *(alt)* with ethylene glycol and 3,3′-(thiazolo[5,4-d]thiazole-2,5-diyl)diphenol			

56

No.	$10^4\,d\bar{v}/dT$ (cc g⁻¹ deg⁻¹)	d_a (g cm⁻³)	d_c (g cm⁻³)	T_g (°C)	T_d (°C)	T_m (°C)	References
C 28							
C 29							
C 30							
C 31	1.2			200	$211_{0.9}$		311, 33, 326, 16, (199)
					$-80_{8.7}$		16
C 32				213			93
C 33				182s			548
C 34				170s			548
C 35						250	170
C 36			(1.17)	123		170	156
C 37			(1.18)	134		222	156, (548)
C 38				179s			548
C 39			(1.38)	173		270	156, 170
C 40				137s		200	170, (548)
C 41			(1.20)	175		260	156, (548)
C 42			(1.21)	167		250	156, (548)
C 43						110	170
C 44					135_1		272
					-10_1		272
C 45			(1.14)			220	156
C 46						226	361
C 47						156	361
C 48						213	361
C 49			(1.27)	121		230	156
C 50						300	156
C 51						275	361

No.	Polymer	$\dfrac{\delta}{\text{cal}^{1/2}\,\text{cm}^{-3/2}}$	$\dfrac{\Delta H_u}{\text{cal mol}^{-1}}$	n
C 52	**Carbonic acid,** polyester with 4,4'-ethylidenediphenol			1.5937
C 53	—, polyester with 4,4'-(1-ethylpropylidene)diphenol			
C 54	—, polyester with 4,4'-isobutylidenediphenol			1.5702
C 55	—, polyester with 4,4'-isopropylidenebis(2-chloro-6-methylphenol)			
C 56	—, polyester with 4,4'-isopropylidenebis(2-chlorophenol)			1.5900
C 57	—, polyester with 4,4'-isopropylidenebis(2,6-dibromophenol)			1.6147
C 58	—, polyester with 4,4'-isopropylidenebis(2,6-dichlorophenol)			1.6056
C 59	—, polyester with 4,4'-isopropylidenebis(2-isopropylphenol)			
C 60	—, polyester with 4,4'-isopropylidenedi-o-cresol			1.5783
C 61	—, polyester with 4,4'-isopropylidenediphenol		8800	1.5850
C 62	—, polyester with 4,4'-(1-isopropylisobutylidene)diphenol			
C 63	—, polyester with 4,4'-(α-methylbenzylidene)diphenol			1.6130
C 64	—, polyester with 4,4'-(1-methylbutylidene)diphenol			1.5745
C 65	—, polyester with 4,4'-methylenedi-o-cresol			
C 66	—, polyester with 4,4'-methylenediphenol			
C 67	—, polyester with 4,4'-(1-methyloctylidene)diphenol			
C 68	—, polyester with 4,4'-(1-methylpentylidene)diphenol			
C 69	—, polyester with 4,4'-oxydiphenol			
C 70	—, polyester with phenolphthalein			
C 71	—, polyester with 4,4'-(p-phenylenedioxy)diphenol			
C 72	—, polyester with 4,4'-(1-propylbutylidene)diphenol			1.5602
C 73	—, polyester with 4,4'-sulfinyldiphenol			
C 74	—, polyester with 4,4'-sulfonyldi-o-cresol			
C 75	—, polyester with 4,4'-sulfonyldiphenol			
C 76	—, polyester with cis-2,2,4,4-tetramethyl-1,3-cyclobutanediol			

No.	$\dfrac{10^4 \, d\bar{v}/dT}{\text{cc g}^{-1} \text{deg}^{-1}}$	$\dfrac{d_a}{\text{g cm}^{-3}}$	$\dfrac{d_c}{\text{g cm}^{-3}}$	$\dfrac{T_g}{°C}$	$\dfrac{T_d}{°C}$	$\dfrac{T_m}{°C}$	References
C 52			(1.22)	130		195	156
C 53				149s		195	170
C 54			(1.18)	149		180	156
C 55				154s			548
C 56			(1.32)	147		210	156, (548)
C 57			(1.91)	157		260	156
C 58			(1.42)	180	240_1	260	156, 396, 198, (548)
					75_1		198
					-50_{1000}		396
C 59						110	170
C 60			(1.22)	95	155_1	170	156, 198, (548)
					-120_1		198
C 61	2.41 / 4.81	(1.20)	1.30	150	150_1	267	552, 156, 431, 554, 198, 66
	(2.0)/(5.9)		1.315		80_{1000}		553, 396, (387), (548)
					-110_1		198
C 62				142s		220	170
C 63			(1.21)	176	190_1	230	156, 198
					-120_1		198
C 64			(1.13)	137		220	156
C 65				51s			548
C 66		1.240	1.303	120		240	554, (156)
C 67						190	156
C 68						200	156
C 69						235	156
C 70				240			362
C 71						215	170
C 72			(1.16)	148		200	156
C 73						250	156
C 74						310	170
C 75						210	156
C 76			1.095		160_{100}	253	357, 224

No.	Polymer	$\dfrac{\delta}{\mathrm{cal}^{1/2}\,\mathrm{cm}^{-3/2}}$	$\dfrac{\Delta H_u}{\mathrm{cal\ mol^{-1}}}$	n
C 77	**Carbonic acid,** polyester with *trans*-2,2,4,4-tetramethyl-1,3-cyclobutanediol			
C 78	—, polyester with 3,3,3′,3′-tetramethyl-1,1′-spirobi[indan]-6,6′-diol			
C 79	—, polyester with 4,4′-(thiazolo[5,4-*d*]thiazole-2,5-diyl)bis(2-methoxyphenol)			
C 80	—, polyester with 3,3′-(thiazolo[5,4-d]thiazole-2,5-diyl)diphenol			
C 81	—, polyester with 4,4′-(thiazolo[5,4-d]thiazole-2,5-diyl)diphenol			
C 82	—, polyester with 4,4′-thiodiphenol			
C 83	—, polyester with 4,4′-[2,2,2-trifluoro-1-(trifluoromethyl)ethylidene]diphenol			
C 84	—, polyurea *(alt)* with 1,4-butanediamine and 1,6-hexanediamine			
C 85	—, polyurea *(alt)* with 1,4-butanediamine and 1,8-octanediamine			
C 86	—, polyurea with 1,10-decanediamine			
C 87	—, polyurea *(alt)* with 1,10-decanediamine and 1,12-dodecanediamine			
C 88	—, polyurea *(alt)* with 1,10-decanediamine and 1,6-hexanediamine			
C 89	—, polyurea *(alt)* with 1,10-decanediamine and 4,4′-methylenedianiline			
C 90	—, polyurea *(alt)* with 1,10-decanediamine and 1,18-octadecanediamine			
C 91	—, polyurea *(alt)* with 1,10-decanediamine and 1,8-octanediamine			
C 92	—, polyurea *(alt)* with 1,10-decanediamine and 1,14-tetradecanediamine			
C 93	—, polyurea *(alt)* with 1,10-decanediamine and *p*-xylene-α,α′-diamine			
C 94	—, polyurea with 1,12-dodecanediamine			
C 95	—, polyurea *(alt)* with 1,12-dodecanediamine and 1,6-hexanediamine			
C 96	—, polyurea *(alt)* with 1,12-dodecanediamine and 4,4′-methylenedianiline			
C 97	—, polyurea *(alt)* with 1,12-dodecanediamine and *p*-xylene-α,α′-diamine			
C 98	—, polyurea *(alt)* with ethylenediamine and 1,6-hexanediamine			
C 99	—, polyurea *(alt)* with 1,7-heptanediamine and 1,6-hexanediamine			
C 100	—, polyurea with 1,6-hexanediamine			
C 101	—, polyurea *(alt)* with 1,6-hexanediamine and 4,4′-methylenedianiline			
C 102	—, polyurea *(alt)* with 1,6-hexanediamine and 1,9-nonanediamine			
C 103	—, polyurea *(alt)* with 1,6-hexanediamine and 1,18-octadecanediamine			
C 104	—, polyurea *(alt)* with 1,6-hexanediamine and 1,8-octanediamine			
C 105	—, polyurea *(alt)* with 1,6-hexanediamine and 1,5-pentanediamine			

No.	$10^4\ d\bar{v}/dT$ cc g^{-1} deg^{-1}	d_a g cm^{-3}	d_c g cm^{-3}	T_g °C	T_d °C	T_m °C	References
C 77			1.08		160_{100}	>360	357, 224
			(1.1)				357
C 78					$240_{0.1}$		455
					-100_1		455
C 79						295	361
C 80						360	361
C 81						332	361
C 82		1.355	1.500	110		240	554, 156
C 83						400	170
C 84						283	150
C 85						220	150
C 86						250	150, (157), (559)
C 87						190	559
C 88						216	559, (150)
C 89				46		246	559
C 90						172	559
C 91						201	559
C 92				40		180	559
C 93				55		260	559
C 94						205	150
C 95				49		241	559, 150
C 96				51		240	559
C 97						258	559
C 98						293	150
C 99						243	150
C 100						300	150, 157, (559)
C 101				55		274	559
C 102						243	150
C 103						192	559
C 104						253	150, (559)
C 105						251	150

No.	Polymer	$\dfrac{\delta}{\text{cal}^{1/2}\,\text{cm}^{-3/2}}$	$\dfrac{\Delta H_u}{\text{cal mol}^{-1}}$	n
C 106	**Carbonic acid,** polyurea *(alt)* with 1,6-hexanediamine and 1,3-propanediamine			
C 107	—, polyurea *(alt)* with 1,6-hexanediamine and 1,14-tetradecanediamine			
C 108	—, polyurea *(alt)* with 1,6-hexanediamine and p-xylene-α,α′-diamine			
C 109	—, polyurea *(alt)* with 4,4′-methylenedianiline and 1,18-octadecanediamine			
C 110	—, polyurea *(alt)* with 4,4′-methylenedianiline and 1,8-octanediamine			
C 111	—, polyurea *(alt)* with 4,4′-methylenedianiline and 1,14-tetradecanediamine			
C 112	—, polyurea with 1,9-nonanediamine			
C 113	—, polyurea *(alt)* with 1,18-octadecanediamine and p-xylene-α,α′-diamine			
C 114	—, polyurea with 1,8-octanediamine			
C 115	—, polyurea *(alt)* with 1,8-octanediamine and p-xylene-α,α′-diamine			
C 116	—, polyurea with 3,3′-(phenylphosphinidene)bis(propylamine)			
C 117	—, polyurea *(alt)* with 1,14-tetradecanediamine and p-xylene-α,α′-diamine			
C 118	**Carbon Monoxide.** See **Ethylene,** polyketone *(alt)* with carbon monoxide			
C 119	**Cellulose**			
C 120	—, ethyl ether	10.3		
C 121	**Cellulose acetate**	10.9		
		11.35		
C 122	**Cellulose butyrate**			3000
C 123	**Cellulose decanoate**			
C 124	**Cellulose heptanoate**			
C 125	**Cellulose hexanoate**			
C 126	**Cellulose laurate**			
C 127	**Cellulose myristate**			
C 128	**Cellulose nitrate**	10.56	1500	
		10.48		

No.	$\dfrac{10^4\,d\bar{v}/dT}{\text{cc g}^{-1}\,\text{deg}^{-1}}$	$\dfrac{d_a}{\text{g cm}^{-3}}$	$\dfrac{d_c}{\text{g cm}^{-3}}$	$\dfrac{T_g}{{}^\circ\text{C}}$	$\dfrac{T_d}{{}^\circ\text{C}}$	$\dfrac{T_m}{{}^\circ\text{C}}$	References
C 106						266	150
C 107				47		198	559
C 108				72		306	559
C 109				48		225	559
C 110				50		263	559
C 111						229	559
C 112			$(1.061)^{25}$		115_{74}	215	162, 150
					-5_{116}		162
					-120_{171}		162
C 113				52		226	559
C 114						260	157, 150
C 115				68		278	559
C 116		1.174					153
C 117				54		248	559
C 118							
C 119		(1.56)	1.625 I			>270	33, 194
			1.62 II				195
			1.61 IV				195
			1.615 X				195
C 120	3/		(1.15)	43		165	2, 33
C 121	2.34/2.70	1.28	1.30	157	175_1	306	11, 121, 370, 363, 55
	2.04/2.34			112	-48_1		7, 121, 363, (4)
	1.85/2.04			46			121
C 122	3.70/4.85			120		207	4, (55)
	3/3.70			40			4
C 123						88	55
C 124						88	55
C 125						94	55
C 126						91	55
C 127						106	55
C 128	3/	(1.6)	2.07				11, 26, 33, 66

No.	Polymer	$\dfrac{\delta}{\text{cal}^{1/2}\,\text{cm}^{-3/2}}$	$\dfrac{\varDelta H_u}{\text{cal mol}^{-1}}$	n
C 129	**Cellulose octanoate**		3100	
C 130	**Cellulose palmitate**			
C 131	**Cellulose propionate**			
C 132	**Cellulose valerate**			
C 133	**Chloral,** polyacetal			
C 134	**Chloroacetic acid.** See Acetic acid, chloro-.			
C 135	**Chloroprene**	9.38	2000	1.558^{20}
		9.25		
		8.6		
C 136	−, *cis*-1,4-polymer			
C 137	**Crotonic acid,** *tert*-butyl ester			
C 138	**Cyclobutane, 1,1-dimethyl-**			
C 139	−, **1,2-dimethyl-.** See **2-Butene,** copolymer *(alt)* with ethylene			
C 140	−, **1-ethyl-1-methyl-**			
C 141	**Cyclobutanepropionic acid, 3-amino-2,2-dimethyl-,** polyamide			
C 142	**Cyclohexane, allyl-,** (stereoregular)			
C 143	−, **1-methoxy-2-vinyl-,** (stereoregular)			
C 144	−, **vinyl-,** (isotactic)			
C 145	**Cyclohexaneacetic acid, 3-(aminomethyl)-,** polyamide			
C 146	**Cyclohexanecarboxylic acid,** vinyl ester			
C 147	−, **undecafluoro-,** vinyl ester			1.369^{25}_D
C 148	*trans*-**1,4-Cyclohexanedicarboxylic acid,** polyester with *cis*-1,4-cyclohexanedimethanol			
C 149	−, polyester with *trans*-1,4-cyclohexanedimethanol			
C 150	−, polyester with ethylene glycol			
C 151	−, polyester with 1,3-propanediol			
C 152	−, polyester with *p*-xylene-α,α'-diol			
C 153	**1,4-Cyclohexanedione,** polymercaptole with 2,2-bis(mercaptomethyl)-1,3-propanedithiol			
C 154	**Cyclohexanone,** polymercaptole with 1,6-hexanedithiol			
C 155	**Cyclohexene, 4-vinyl-,** (isotactic)			

No.	$\dfrac{10^4\,d\bar{v}/dT}{\text{cc g}^{-1}\,\text{deg}^{-1}}$	$\dfrac{d_a}{\text{g cm}^{-3}}$	$\dfrac{d_c}{\text{g cm}^{-3}}$	$\dfrac{T_g}{{}^\circ\text{C}}$	$\dfrac{T_d}{{}^\circ\text{C}}$	$\dfrac{T_m}{{}^\circ\text{C}}$	References
C 129						116	364, (55)
C 130						105	55
C 131						234	55
C 132						122	55
C 133						140	355
C 134							
C 135	/5.0	1.243^{25}	1.356^{25}	-45		43	7,2,326,61,393,283,25
							9, (38)
							(172), (13)
C 136		1.283^{25}		-20		70	393
C 137				86	82_1		402
C 138					-20_1		345
C 139							
C 140				-23			457
C 141						358	341
C 142				75s	$75_{0.1}$	230	292, 113, 117, (401)
C 143						195	448
C 144		0.95	0.982	95s	$90_{0.1}$	383	66, 68, 292, 113, 401
		(0.945)	0.95		$-90_{8,400}$		68, 274, 547, (111)
C 145						297	341
C 146					-76_{6490}		547
					-157_{8800}		547
C 147				54			359
C 148						205	384
C 149						246	384
C 150			$(1.219)^\circ$	<25		120	32, 180
C 151						110	384
C 152						106	384
C 153						315	241
C 154						75	241
C 155		(1.128)				418	366

No.	Polymer	$\dfrac{\delta}{\text{cal}^{1/2}\,\text{cm}^{-3/2}}$	$\dfrac{\Delta H_u}{\text{cal mol}^{-1}}$	n
C 156	**Cyclohexene oxide.** See **7-Oxabicyclo[4·1·0]heptane.**			
C 157	**Cyclohexene sulfide.** See **7-Thiabicyclo[4·1·0]heptane.**			
C 158	**Cyclopentane, allyl-,** (stereoregular)			
C 159	—, **vinyl-,** (stereoregular)			
C 160	—, 1,3-*trans*-vinylene polymer *(alt)*			
C 161	**Cyclopentene,** copolymer *(alt)* with ethylene			
C 162	**Cyclopropane, 1,1-dimethyl-**			
C 163	—, **vinyl-,** (isotactic)			
C 164	*cis*-**1,2-Cyclopropanedicarboxylic acid,** polyamide with 1,6-hexanediamine			
C 165	—, polyester with hydroquinone			
C 166	—, polyester with 4,4′-isopropylidenediphenol			
C 167	—, polyester with resorcinol			
C 168	*trans*-**1,2-Cyclopropanedicarboxylic acid,** polyamide with ethylenediamine			
C 169	—, polyamide with 1,6-hexanediamine			
C 170	—, polyamide with piperazine			
C 171	—, polyamide with 1,2-propanediamine			
C 172	—, polyester with hydroquinone			
C 173	—, polyester with 4,4′-isopropylidenediphenol			
C 174	—, polyester with resorcinol			
C 175	*trans*-**1,2-Cyclopropanedicarboxylic acid, 1-methyl-,** polyamide with 1,6-hexanediamine			
C 176	—, —, polyamide with piperazine			
C 177	—, —, polyester with 4,4′-isopropylidenediphenol			
C 178	—, **3-methyl-,** polyamide with 1,6-hexanediamine			
C 179	—, —, polyamide with piperazine			
C 180	—, —, polyester with 4,4′-isopropylidenediphenol			
C 181	**Cyclopropanone,** polyketone. See **Ethylene,** polyketone *(alt)* with carbon monoxide.			
C 182	**Cyclotetrasiloxane.** See **Silicone.**			
C 183	**Cyclotrisilazane.** See **Phosphonitrile.**			

No.	$\dfrac{10^4 \, d\bar{v}/dT}{cc \, g^{-1} \, deg^{-1}}$	d_a g cm^{-3}	d_c g cm^{-3}	T_g °C	T_d °C	T_m °C	References
C 156							
C 157							
C 158					$60_{0.1}$	225	113, 117, (401)
C 159		(0.965)	0.986		$75_{0.1}$	292	68, 113, 401
C 160						202	142
C 161		$(1.01)^{24}$	1.03			185	228
C 162		(0.872)		26	-10_1	66α	221, 390, 345
				-12		60.5γ	221, 390
						48.5β	390
C 163						230	194
C 164						180	340
C 165						160	340
C 166						130	340
C 167						65	340
C 168						350	340
C 169						300	340
C 170						330	340
C 171						310	340
C 172						280	340
C 173						180	340
C 174						105	340
C 175						115	340
C 176						130	340
C 177						90	340
C 178						270	340
C 179						280	340
C 180						130	340
C 181							
C 182							
C 183							

No.	Polymer	$\dfrac{\delta}{\mathrm{cal}^{1/2}\ \mathrm{cm}^{-3/2}}$	$\dfrac{\varDelta\,H_u}{\mathrm{cal}\ \mathrm{mol}^{-1}}$	n
D 1	**Decanedioic acid.** See **Sebacic acid.**			
D 2	**Decanoic acid, 10-amino-**, polyamide			
D 3	−, **10-hydroxy-**, polyester			
D 4	−, **nonadecafluoro-**, vinyl ester			1.334_D^{25}
D 5	**1-Decanol, 10,10′-hexamethylenedioxydi-**, polyether			
D 6	−, **10,10′-pentamethylenedioxydi-**, polyether			
D 7	−, **10,10′-tetramethylenedioxydi-**, polyether			
D 8	**1-Decene,** (isotactic)			
D 9	**Diglycolic acid,** polyamide with 1,6-hexanediamine			
D 10	−, polyamide with 1,5-pentanediamine			
D 11	−, polyamide with piperazine			
D 12	**Diglycolimide, N-vinyl-**			
D 13	**2-m-Dioxaneëthanol,** $\beta,\beta,5,5$-**tetramethyl-**, methacrylate. See **Methacrylic acid,** 2,2-dimethyl-2-(5,5-dimethyl-m-dioxan-2-yl)ethyl ester.			
D 14	**5-m-Dioxanemethanol, 5-methyl-**, methacrylate. See **Methacrylic acid,** 5-methyl-m-dioxan-5-ylmethyl ester.			
D 15	−, **2,2,5-trimethyl-**, methacrylate. See **Methacrylic acid,** 2,2,5-trimethyl-m-dioxan-5-ylmethyl ester.			
D 16	**p-Dioxin, 2,3-dihydro-**			
D 17	**Diphenic acid,** polyamide with fluorene-9,9-bis(propylamine)			
D 18	−, polyamide with 1,6-hexanediamine			
D 19	−, polyamide with piperazine			
D 20	−, polyamide with m-xylene-α,α'-diamine			
D 21	−, polyester with ethylene glycol			
D 22	**Docosanedioic acid,** polyamide with 1,10-decanediamine			
D 23	−, polyamide with 1,12-dodecanediamine			
D 24	−, polyamide with 1,6-hexanediamine			
D 25	−, polyamide with 1,18-octadecanediamine			

No.	$\dfrac{10^4\,d\bar{v}/dT}{\text{cc g}^{-1}\,\text{deg}^{-1}}$	$\dfrac{d_a}{\text{g cm}^{-3}}$	$\dfrac{d_c}{\text{g cm}^{-3}}$	$\dfrac{T_g}{°C}$	$\dfrac{T_d}{°C}$	$\dfrac{T_m}{°C}$	References
D 1							
D 2	(3.5)/(4.2)	$(1.032)^{25}$	1.019	43	94_{70}	192	106, 162, 195
		$(1.051)^{30}$			-39_{129}		106, 162, (321), (348)
					-120_{179}		162, (350)
D 3						76	3
D 4				−18			359
D 5						76.5	540
D 6						72.3	540
D 7						73.2	540
D 8					-35_{180}	40	97, 419, (372)
					-160_{590}		97
D 9						172	148
D 10						130	157
D 11						258	148
D 12				100s			199
D 13							
D 14							
D 15							
D 16						280	86
D 17				165s			432
D 18				127s			432
D 19				193s			432
D 20				159s			432
D 21				132s			180
D 22			$(1.010)^{25}$			169	529
D 23			$(1.006)^{25}$			164	529
D 24			$(1.032)^{25}$			180	529
D 25			$(0.990)^{25}$			146	529

No.	Polymer	$\dfrac{\delta}{\mathrm{cal}^{1/2}\ \mathrm{cm}^{-3/2}}$	$\dfrac{\Delta H_u}{\mathrm{cal\ mol}^{-1}}$	n
D 26	**Docosanedioic acid,** polyamide with 1,8-octanediamine			
D 27	—, polyamide with 2,2'-*p*-phenylenebis(ethylamine)			
D 28	—, polyamide with 1,14-tetradecanediamine			
D 29	—, polyamide with *p*-xylene-α,α'-diamine			
D 30	**Docosanoic acid, 22-amino-,** polyamide			
D 31	**Dodecanedioic acid,** polyamide with *cis*-1,4-cyclohexanebis(methylamine)			
D 32	—, polyamide with *trans*-1,4-cyclohexanebis(methylamine)			
D 33	—, polyamide with 1,10-decanediamine			
D 34	—, polyamide with 1,12-dodecanediamine			
D 35	—, polyamide with ethylenediamine			
D 36	—, polyamide with 1,6-hexanediamine			
D 37	—, polyamide with 1,18-octadecanediamine			
D 38	—, polyamide with 1,8-octanediamine			
D 39	—, polyamide with 2,2'-*p*-phenylenebis(ethylamine)			
D 40	—, polyamide with 1,14-tetradecanediamine			
D 41	—, polyamide with *p*-xylene-α,α'-diamine			
D 42	—, polyester with *cis*-1,4-cyclohexanedimethanol			
D 43	—, polyester with *trans*-1,4-cyclohexanedimethanol			
D 44	—, polyester with diethylene glycol			
D 45	—, polyester with 1,8-octanediol			
D 46	—, polyester with *p*-xylene-α,α'-diol			
D 47	**2,11-Dodecanedione,** polymercaptole with 2,2-bis(mercaptomethyl)-1,3-propanedithiol			
D 48	**Dodecanoic acid.** See **Lauric acid** (unsubstituted only).			
D 49	—, **12-amino-,** polyamide			
D 50	**1-Dodecene,** (isotactic)			

No.	$\dfrac{10^4\,d\bar{v}/dT}{cc\,g^{-1}\,deg^{-1}}$	$\dfrac{d_a}{g\,cm^{-3}}$	$\dfrac{d_c}{g\,cm^{-3}}$	$\dfrac{T_g}{°C}$	$\dfrac{T_d}{°C}$	$\dfrac{T_m}{°C}$	References
D 26			$(1.021)^{25}$			175	529
D 27			$(1.050)^{25}$			230	529
D 28			$(0.999)^{25}$			153	529
D 29			$(1.056)^{25}$			225	529
D 30						145	106
D 31						215	410
D 32						278	410
D 33			$(1.057)^{25}$			192	529, (354), (348)
D 34			$(1.036)^{25}$			183	529
D 35						261s	348
D 36			$(1.070)^{25}$		$56_{3.8}$	219	529, 16, (348)
					$-50_{5.3}$		16
					$-120_{6.2}$		16
D 37			$(1.015)^{25}$			170	529
D 38			$(1.062)^{25}$			200	529, 348
D 39			$(1.108)^{25}$			280	529
D 40			$(1.027)^{25}$			175	529
D 41			$(1.129)^{25}$	105	105_{110}	272	529, 410
D 42						46	384
D 43						85	384
D 44				-71			51
D 45						73	157
D 46						94	384
D 47						160	241
D 48							
D 49 (3.8/4.5)			$(1.029)^{30}$	37		179	106, 348
D 50				-36	-6_{150}	49_{I}	76, 97, 3/2
					-145_{520}	-25_{II}	97, 46

No.	Polymer	$\dfrac{\delta}{\mathrm{cal}^{1/2}\ \mathrm{cm}^{-3/2}}$	$\dfrac{\Delta H_u}{\mathrm{cal\ mol}^{-1}}$	n
E 1	**Eicosanedioic acid,** polyamide with 1,10-decanediamine			
E 2	—, polyamide with 1,6-hexanediamine			
E 3	—, polyamide with 1,8-octanediamine			
E 4	**Enanthaldehyde. See Heptanal.**			
E 5	**Enanthic acid. See Heptanoic acid.**			
E 6	**Epibromohydrin,** polyether			
E 7	**Epichlorohydrin,** polyether (isotactic)	9.4		
E 8	**Epifluorohydrin,** polyether			
E 9	**Ethanedioic acid. See Oxalic acid.**			
E 10	**Ether, allyl glycidyl. See Propane, 1-(allyloxy)-2,3-epoxy-.**			
E 11	**—, benzyl vinyl,** (isotactic)			
E 12	**—, 2-butoxyethyl _p_-vinylbenzyl**			
E 13	**—, 4-butylcyclohexyl vinyl**			
E 14	**—, butyl vinyl**			1.4563_D^{30}
E 15	**—, butyl vinyl**			
E 16	**—, _sec_-butyl vinyl,** (isotactic)			1.4740_D^{30}
E 17	**—, _tert_-butyl vinyl,** (isotactic)			
E 18	**—, butyl _p_-vinylbenzyl**			
E 19	**—, _sec_-butyl _p_-vinylbenzyl**			
E 20	**—, 2-chloroethyl vinyl,** (isotactic)			
E 21	**—, cyclohexyl vinyl**			
E 22	**—, decahydro-2-naphthyl vinyl**			
E 23	**—, decyl vinyl**			1.4628_D^{30}
E 24	**—, 2,2-dimethylbutyl vinyl**			
E 25	**—, dodecyl vinyl**			1.4640_D^{30}
E 26	**—, 2-ethoxyethyl _p_-vinylbenzyl**			
E 27	**—, 2-ethylhexyl vinyl**			1.4626_D^{30}
E 28	**—, 2-ethylhexyl _p_-vinylbenzyl**			

No.	$10^4 \, d\bar{v}/dT$ $\mathrm{cc \ g^{-1} \ deg^{-1}}$	d_a $\mathrm{g \ cm^{-3}}$	d_c $\mathrm{g \ cm^{-3}}$	T_g °C	T_d °C	T_m °C	References
E 1						171s	348
E 2						189s	348
E 3						179s	348
E 4							
E 5							
E 6						112	78
E 7	/5.6	1.37^{25}				121	426, (105), (143)
E 8						68	105
E 9							
E 10							
E 11						162	244
E 12					-38_1		196
E 13				74s			199
E 14		0.927^{27}		-55	$-32_{0.8}$		403, 16, (89)
					-70_8		16
					$-150_{10.4}$		16
E 15		(0.93)	0.943			64	179, 244
E 16		0.924^{30}		-20			403
E 17					$83_{1.7}$	260	16, 244, (293)
E 18					10_1		196
E 19					40_1		196
E 20						150	244
E 21				62s	$-1_{7,\,240}$		199, 547
					$-168_{9,\,800}$		547
E 22			(1.037)	29b			199
E 23		0.883^{30}				7	403
E 24				9s			183
E 25		0.892^{30}				30	403
E 26					0_1		196
E 27		0.904^{27}		-66			403
E 28					-23_1		196

No.	Polymer	$\dfrac{\delta}{\text{cal}^{1/2}\,\text{cm}^{-3/2}}$	$\dfrac{\Delta H_u}{\text{cal mol}^{-1}}$	n
E 29	**Ether, ethyl propenyl,** (stereoregular)			
E 30	**−, ethyl vinyl**			1.4540_D^{30}
E 31	**−, ethyl vinyl,** (isotactic)			
E 32	**−, glycidyl phenyl.** See **Propane, 1,2-epoxy-3-phenoxy.**			
E 33	**−, hexyl vinyl**			1.4591_D^{30}
E 34	**−, hexyl *p*-vinylbenzyl**			
E 35	**−, 4-hydroxybutyl *p*-vinylbenzyl**			
E 36	**−, 2-hydroxyethyl *p*-vinylbenzyl**			
E 37	**−, isobutyl vinyl**			1.4507_D^{30}
E 38	**−, −,** (isotactic)			
E 39	**−, isopropenyl methyl**			
E 40	**−, isopropyl vinyl**			
E 41	**−, −,** (isotactic)			
E 42	**−, 2-methoxyethyl vinyl,** (isotactic)			
E 43	**−, methyl propenyl,** (stereoregular)			
E 44	**−, methyl vinyl**			1.467_D^{20}
E 45	**−, −,** (isotactic)			1.4700_D^{30}
E 46	**−, methyl *p*-vinylbenzyl**			
E 47	**−, neopentyl vinyl,** (isotactic)			
E 48	**−, octyl vinyl**			1.4613_D^{30}
E 49	**−, octyl *p*-vinylbenzyl**			
E 50	**−, pentyl vinyl**			1.4581_D^{30}
E 51	**−, phenyl *p*-vinylphenyl**			
E 52	**−, propenyl propyl,** (stereoregular)			
E 53	**−, propyl vinyl**			

No.	$\dfrac{10^4\ d\bar{v}/dT}{cc\ g^{-1}\ deg^{-1}}$	$\dfrac{d_a}{g\ cm^{-3}}$	$\dfrac{d_c}{g\ cm^{-3}}$	$\dfrac{T_g}{°C}$	$\dfrac{T_d}{°C}$	$\dfrac{T_m}{°C}$	References
E 29						231	226, (245)
E 30		0.968^{27}		-42	$-30_{0.1}$		403, 532, (56)
		0.944^{20}			$-173_{0.90}$		314, (16), (199)
		0.976^{20}					532
E 31						86	244
E 32							
E 33		0.925^{27}		-77			403
E 34					-20_1		196
E 35					-20_1		196
E 36					46_1		196
E 37		0.93		-19	$-1_{1.2}$		403, 89, 16, (199)
		0.916^{30}			$-150_{11.8}$		403, 16, (183), (56)
E 38		$(0.91)^{20}$	0.940	-18		170	199, 66, 89, 293,
							(159), (244)
E 39				67			403
E 40		0.924^{30}		-3			403
E 41			$(0.93)^{30}$			191	159, 293, (244)
E 42						73	244
E 43						287	226, 245
E 44				$-31s$	$-10_{27.2}$		326, 183, 16
E 45		$(1.037)^{27}$	$(1.092)^{23}$	-21		150	403, 293, 76, (244)
E 46					77_1		196
E 47			(0.91)			216	159, 244, 293
E 48		0.914^{27}		-79			403
E 49					-42_1		196
E 50		0.918^{30}		-66			403
E 51				$100s$			96
E 52						168	226, 245
E 53					$-27_{1.1}$		16
					$-70_{4.2}$		16
					$-160_{5.5}$		16

No.	Polymer	$\dfrac{\delta}{cal^{1/2}\ cm^{-3/2}}$	$\dfrac{\Delta H_u}{cal\ mol^{-1}}$	n
E 54	**Ether, propyl vinyl,** (isotactic)			
E 55	**–, propyl *p*-vinylbenzyl**			
E 56	**–, 2,2,2-trifluoroethyl trifluorovinyl**			
E 57	**–, 2,2,2-trifluoroethyl vinyl,** (isotactic)			
E 58	**Ethyl cellulose.** See **Cellulose,** ethyl ether.			
E 59	**Ethylene**	7.9	1920	
		8.12	1850	
		8.36	1820	
E 60	**Ethylene,** polyketone *(alt)* with carbon monoxide			
E 61	**–, bromo-**	9.6		
E 62	**–, chloro-**	9.48	656	1.539_D^{20}
		9.53		
		9.55		
E 63	**–, chlorotrifluoro-**	7.2	1200	
		7.87	1300	
			2108	
E 64	**–, 1,1-dibromo-**			
E 65	**–, 1,1-dichloro-**	12.2	(330)	1.60_D^{20}
E 66	**–, 1,1-difluoro-**			1.42^{25}
E 67	**–, 1,2-difluoro-**			
E 68	**–, fluoro-**		1800	
E 69	**–, tetrafluoro-**	6.21	1370	

---

* Crystalline second-order transition.
** Extrapolated value.

No.	$10^4\, d\bar{v}/dT$ cc g⁻¹ deg⁻¹	d_a g cm⁻³	d_c g cm⁻³	T_g °C	T_d °C	T_m °C	References
E 54						76	244
E 55					22_1		196
E 56				35			458
E 57					128		244
E 58							
E 59	1.0/8.8	0.855^{30}	0.991^{30}	−20	$-5_{4.1}$	141.4 I	8, 165, 91, 288, 287
	/8.92	0.887^{25}	1.001^{25}	−120	$-107_{8.6}$	95 II	64, 16, 165, 7, 567, 568
	/9.59	0.855^{20}	0.990^{20}			−120*	72, 418, 168, 261, 569, 570
E 60		(1.203)	1.296			185	203
E 61							7
E 62	1.41/4.22	1.385^{30}	1.44	85	$90_{0.67}$	285**	10, 83, 326, 514,
	1.11/1.41	$(1.386)^{26}$	1.455	−26	$-30_{8.5}$		534, 60, 16
					-255_{7200}		2, 407, 276, (22)
							7, (11), (513)
E 63	(1.0)/(2.0)	2.11^{20}	2.25^{25}	52	90_1	218	35, 88, 38, 231, 163
	1.492/2.505	2.077^{20}	2.187^{20}		-20_1		169, 23, 309
	/3.47	2.032^{25}	2.186		$-58_{10,000}$		67, (16), (130), (30)
E 64		$(2.83)^{18}$	3.065				279
E 65	(2.37)/(4.78)	1.66^{25}	1.95^{25}	−18	$80_{5.5}$	210	2, 278, 199, 254, 38
			1.959		15_{11}		279, 16, 3, 279
			1.946				278
E 66	/(2.1)	(1.74)	2.00 I	13		171	392, 88, 519, 433
	/(4.6)		1.90 II	−35			392, 519, 88, (555)
E 67				50s			464
E 68		(1.38)	1.44		$41_{1.7}$	200	439, 137, 73, 16
					$-20_{5.4}$		16
E 69	3.0/4.77	2.00	2.304^{23}	130	$127_{0.62}$	327 I	7, 161, 429, 395, 48
			2.42^{25}	−113	$-97_{0.60}$	30	23, 38, 49
						19 II	161

No.	Polymer	$\dfrac{\delta}{\text{cal}^{1/2}\,\text{cm}^{-3/2}}$	$\dfrac{\Delta H_u}{\text{cal mol}^{-1}}$	n
E 70	**Ethylene glycol,** benzoate methacrylate. See **Methacrylic acid,** 2-hydroxyethyl ester, benzoate.			
E 71	—, isobutyrate methacrylate. See **Methacrylic acid,** 2-hydroxyethyl ester, isobutyrate.			
E 72	**Ethylene oxide,** polyether		2000	1.4539_D^{75}
			2220	
E 73	**Ethylene sulfide,** poly(thioether)			
E 74	—, poly(thioether) *(alt)* with hexamethylene sulfide			
E 75	—, poly(thioether) *(alt)* with tetrahydrothiophene			
E 76	**Ethylidene.** See **2-Butene**			
F 1	**Fluorenc-9,9-dipropionic acid,** polyamide with fluorene-9,9-bis(propylamine)			
F 2	—, polyamide with 1,6-hexanediamine			
F 3	—, polyamide with piperazine			
F 4	—, polyamide with m-xylene-α,α'-diamine			
F 5	**Formaldehyde,** polyacetal	10.2	1780	1.510_D^{25}
		11.0	1590	
			1760	
F 6	—, polyacetal with *trans*-1,4-cyclohexanedimethanol			
F 7	—, polyacetal with *trans*-1,4-cyclohexanediol			
F 8	—, polyacetal with 2,5(or 2,6)-norbornanediol			
F 9	—, polyacetal with poly(vinyl alcohol)	9.95		1.50
F 10	—, polyacetal with 2,2,4,4-tetramethyl-1,3-cyclobutanediol			1.480 [27]
F 11	—, polyacetal with *cis*-2,2,4,4-tetramethyl-1,3-cyclobutanediol			

No.	$10^4\ d\bar{v}/dT$ $cc\ g^{-1}\ deg^{-1}$	d_a $g\ cm^{-3}$	d_c $g\ cm^{-3}$	T_g °C	T_d °C	T_m °C	References
E 70							
E 71							
E 72	/6.2	1.13	1.33	−67	$−60_{0.01}$	66.2	220, 453, 101, 267, 265
	/6.6	1.077^{70}	1.205				532, 338, 463, 195
		1.080^{75}	1.23				453, 533
		1.070^{100}					37
E 73			1.22			210	485, (55), (120)
E 74						86	55
E 75						89	120
E 76							
F 1				165s			432
F 2				122s			432
F 3				145s			432
F 4				150s			432
F 5	(1.8)/(2.4)	1.25	1.49 I	−30	$115_{0.15}$	182.5I	123, 23, 571, 5, 298, 553, 166
		1.328^{25}	1.54II	−82	$−75_{0.176}$	60II	118, 416, 532, 551, (27), (45) 570, (152)
F 6						78	230
F 7						209	230
F 8						83	230
F 9	2.4/	1.2		110s			171, 199, 33
	2.3/	1.23					199
F 10		(1.04)	(1.07)			262 I	230
						177II	230
F 11						285 I	230
						213II	230

No.	Polymer	$\dfrac{\delta}{\mathrm{cal}^{1/2}\,\mathrm{cm}^{-3/2}}$	$\dfrac{\Delta H_u}{\mathrm{cal\ mol}^{-1}}$	n
F 12	**Formaldehyde,** polyacetal with *trans*-2,2,4,4-tetramethyl-1,3-cyclobutanediol			
F 13	—, polymercaptal with 1,4-butanedithiol			
F 14	—, polymercaptal with 1,6-hexanedithiol			
F 15	—, **difluorothio-,** polymercaptal			
F 16	—, **seleno-,** poly(selenoacetal)			
F 17	—, **thio-,** polymercaptal			
F 18	**Formic acid,** vinyl ester, (isotactic)			
F 19	—, —, (syndiotactic)			1.4757_D^{20}
F 20	**Fumaric acid,** polyester with 4,4'-*sec*-butylidenediphenol			
F 21	—, polyester with 4,4'-cyclohexylidenediphenol			
F 22	—, polyester with 4,4'-isopropylidenediphenol			
F 23	**Furan, tetrahydro-,** polyether		3000	
F 24	—, **2-vinyl-**			1.55_D^{20}
F 25	**2,5-Furandipropionic acid,** polyanhydride			
F 26	—, **tetrahydro-,** polyanhydride			
F 27	**Furfural,** copolymer *(alt)* with dimethylketene.			
	See **Hydracrylic acid, 3-(2-furyl)-2,2-dimethyl-,** polyester.			

No.	$10^4 \, d\bar{v}/dT$ $\text{cc g}^{-1} \text{deg}^{-1}$	d_a g cm^{-3}	d_c g cm^{-3}	T_g °C	T_d °C	T_m °C	References
F 12						260 I	230
						197 II	230
F 13						73	120, (241)
F 14						69	241
F 15					-118_1	35	495
F 16	(2.82)	2.97I				178I	492, 551
			2.985II				551
F 17	(1.52)	1.603				260	492, 319, 501, (120), (241)
F 18	(1.35)	1.49					385
F 19		(1.3476)		37			311, 385
F 20				135s			154
F 21				180s			154
F 22				170s			154
F 23	(1.06)	1.18		-84	-74_1	43	437, 488, 487, 532
	(1.036)[20]	1.09			-140_1		532, 488, (20), (120)
F 24							311
F 25						67	229
F 26						135	229
F 27							

No.	Polymer	$\dfrac{\delta}{\text{cal}^{1/2}\,\text{cm}^{-3/2}}$	$\dfrac{\varDelta H_u}{\text{cal mol}^{-1}}$	n
G 1	L-**Glutamic acid,** 5-benzyl ester, polyamide			
G 2	—, 5-butyl ester, polyamide			
G 3	—, 5-ethyl ester, polyamide			
G 4	—, 5-methyl ester, polyamide			
G 5	—, 5-propyl ester, polyamide			
G 6	**Glutaric acid,** polyamide with *cis*-1,4-cyclohexanebis(methylamine)			
G 7	—, polyamide with *trans*-1,4-cyclohexanebis(methylamine)			
G 8	—, polyamide with 1,5-pentanediamine			
G 9	—, polyester with *trans*-1,4-cyclohexanedimethanol			
G 10	—, polyester with 1,10-decanediol			
G 11	—, polyester with diethylene glycol			
G 12	—, polyester with 1,3-propanediol			
G 13	—, polyester with *p*-xylene-α,α'-diol			
G 14	—, **3,3-dimethyl-,** polyester with 1,10-decanediol			
G 15	—, —, polyester with 2,2-dimethyl-1,3-propanediol			
G 16	**Glutarimide,** N, 3-trimethylene polymer *(alt)*			
G 17	**Glycerol,** 1,3-diacetate-2-methacrylate. See **Methacrylic acid,** ester with 2-hydroxytrimethylene diacetate.			
G 18	**Glycine,** polyamide			
G 19	**Glycolic acid,** ethyl ester, methacrylate. See **Methacrylic acid,** ester with ethyl glycolate.			
G 20	—, polyester			

No.	$\dfrac{10^4\,d\bar{v}/dT}{cc\ g^{-1}\ deg^{-1}}$	$\dfrac{d_a}{g\ cm^{-3}}$	$\dfrac{d_c}{g\ cm^{-3}}$	$\dfrac{T_g}{°C}$	$\dfrac{T_d}{°C}$	$\dfrac{T_m}{°C}$	References
G 1	1.9/4.0	$(1.177)^{20}$	$(1.272)^{25}$	12	$15_{0.18}$		512, 314
					$-130_{0.30}$		314
G 2					-21_{240}		520
G 3					9_{35}		520
					-13_{150}		520
G 4					-18_{200}		520
G 5					-15_{700}		520
G 6						167	410
G 7						290	410
G 8						198	157
G 9						50	384
G 10						60	358
G 11				-47			51
G 12						53	358
G 13						58	384
G 14						4	358
G 15						10	358
G 16				90		283	543
G 17							
G 18		(1.43)	1.54				301, 300
G 19							
G 20		1.60	1.69	38	50_{100}	260	411, 93, 389
			1.700		-40_{100}		389, 411

6*

No.	Polymer	$\dfrac{\delta}{\text{cal}^{1/2}\,\text{cm}^{-3/2}}$	$\dfrac{\Delta H_u}{\text{cal mol}^{-1}}$	n
H 1	**Heptadecanedioic acid,** polyamide with 2,2'-*p*-phenylenebis(ethylamine)			
H 2	—, polyamide with *p*-xylene-α, α'-diamine			
H 3	—, **17-amino-,** polyamide			
H 4	**Heptanal,** polyacetal (isotactic)			
H 5	**Heptanedioic acid. See Pimelic acid.**			
H 6	**Heptanoic acid, 7-amino-,** polyamide			
H 7	—, **7-(4-carboxybutoxy)-,** polyamide with 1,6-hexanediamine			
H 8	—, —, polyamide with 5,5'-oxybis(pentylamine)			
H 9	—, —, polyamide with *p*-xylene-α, α'-diamine			
H 10	—, **7-(*p*-carboxyphenoxy)-,** polyester with ethylene glycol			
H 11	—, **7-(3-carboxypropoxy)-,** polyamide with 1,6-hexanediamine			
H 12	—, —, polyamide with 5,5'-oxybis(pentylamine)			
H 13	—, —, polyamide with *p*-xylene-α, α'-diamine			
H 14	—, **7,7'-oxydi-,** polyamide with 5,5'-oxybis(pentylamine)			
H 15	—, —, polyamide with *p*-xylene-α, α'-diamine			
H 16	—, **7,7'-thiodi-,** polyamide with *p*-xylene-α, α'-diamine			
H 17	**1-Heptene,** (isotactic)			
H 18	—, **6,6-dimethyl-,** (stereoregular)			
H 19	—, **5-methyl-,** (stereoregular)			
H 20	—, **6-methyl-**			
H 21	**Hexadecanedioic acid,** polyamide with N,N'-dimethyl-*p*-xylene-α, α'-diamine			
H 22	—, polyamide with 2,2'-*p*-phenylenebis(ethylamine)			
H 23	—, polyamide with *p*-xylene-α, α'-diamine			
H 24	—, polyester with diethylene glycol			
H 25	**Hexadecanoic acid. See Palmitic acid.**			
H 26	**1-Hexadecene,** (isotactic)			

No.	$\dfrac{10^4\,d\bar{v}/dT}{\text{cc g}^{-1}\,\text{deg}^{-1}}$	$\dfrac{d_a}{\text{g cm}^{-3}}$	$\dfrac{d_c}{\text{g cm}^{-3}}$	$\dfrac{T_g}{°C}$	$\dfrac{T_d}{°C}$	$\dfrac{T_m}{°C}$	References
H 1			$(1.067)^{25}$			249	529
H 2			$(1.080)^{25}$			239	529
H 3						150	350, 354
H 4						150 I	405
						75 II	405
H 5							
H 6	$(3.5)/(4.2)$	$(1.095)^{25}$	1.211	62	50_1	233	106, 162, 195, 321, 348
					-70_1		77, (157), (350)
					-130_1		77
H 7						147	527
H 8						106	527
H 9						207	527
H 10						55	55
H 11						149	527
H 12						102	527
H 13						200	527
H 14						129	527
H 15						229	527
H 16						228	530
H 17				-40	-31_{180}	17	46, 97
					-148_{320}		97
H 18				40s		104	292
H 19						52	195
H 20				-34s			292
H 21						85	558
H 22			$(1.077)^{25}$		85_{110}	258	529
H 23			$(1.092)^{25}$			248	529
H 24				-68		51	
H 25							
H 26					40_{30}	67.5	97, 372
					-125_{200}		97

No.	Polymer	$\dfrac{\delta}{cal^{1/2}\ cm^{-3/2}}$	$\dfrac{\Delta H_u}{cal\ mol^{-1}}$	n
H 27	**1,3-Hexadiene,** *trans*-1,4-polymer (isotactic)			
H 28	**2,4-Hexadiene, 2,5-dimethyl-,** *trans*-1,4-polymer			
H 29	**2,4-Hexadienedioic acid,** diisopropyl ester, *erythro-trans*-2,5-polymer (disyndiotactic)			
H 30	**Hexanedioic acid. See Adipic acid.** For derivatives see below.			
H 31	−, **3-methyl-,** polyamide with 1,6-hexanediamine			
H 32	−, −, polyamide with spiro[3.3]heptane-2,6-diamine			
H 33	−, -, polyester with 4,4′-isopropylidenediphenol			
H 34	−, **5-methyl-2-methylene-,** dimethyl ester			
H 35	**Hexanoic acid, 6-amino-,** polyamide		4210	
			4290	
			5150	
H 36	−, **6-amino-3-methyl-,** polyamide (atactic)			
H 37	−, −, D(-)-, polyamide (isotactic)			
H 38	−, **6-(4-carboxybutoxy)-,** polyamide with 1,6-hexanediamine			
H 39	−, −, polyamide with 5,5′-oxybis(pentylamine)			
H 40	−, −, polyamide with *p*-xylene-α,α'-diamine			
H 41	−, **6-(3-carboxypropoxy)-,** polyamide with 1,6-hexanediamine			
H 42	−, −, polyamide with 5,5′-oxybis(pentylamine)			
H 43	−, −, polyamide with *p*-xylene-α,α'-diamine			
H 44	−, **6-hydroxy-,** polyester			
H 45	−, **6-mercapto-,** poly(thioester)			
H 46	−, **6,6′-oxydi-,** polyamide with 1,6-hexanediamine			
H 47	−, −, polyamide with 5,5′-oxybis(pentylamine)			
H 48	−, −, polyamide with *p*-xylene-α,α'-diamine			
H 49	−, **6,6′-thiodi-,** polyamide with *p*-xylene-α,α'-diamine			
H 50	−, **undecafluoro-,** vinyl ester			1.344_D^{25}
H 51	**1-Hexanol, 6,6′-decamethylenedioxydi-,** polyether			
H 52	−, **6,6′-pentamethylenedioxydi-,** polyether			
H 53	−, **6,6′tetramethylenedioxydi-,** polyether			
H 54	**1-Hexene,** (isotactic)			

No.	$10^4\ d\bar{v}/dT$ cc g⁻¹ deg⁻¹	d_a g cm⁻³	d_c g cm⁻³	T_g °C	T_d °C	T_m °C	References
H 27		(0.97)	1.00				297
H 28						265	136
H 29			1.06			230	459
H 30							
H 31		(1.094)		17		230	557, (157)
H 32						300	509
H 33				56s			371
H 34	2.0/5.6			34			460, (235)
H 35		1.084 0.999^{220} 0.998^{230} 0.996^{245}	1.225α 1.150β 1.155γ		75_5 $-55_{7.1}$ $-120_{8.4}$	228.5	23, 346, 565, 325, (162) 325, 346, 16, (425) 570, (299), (350), (348) 325, (321), (106)
H 36						150	494
H 37		(1.06)	1.11			225	494, 499
H 38						149	527
H 39						100	527
H 40						206	527
H 41						152	527
H 42						108	527
H 43						190	527
H 44						53	3
H 45						106	374
H 46						175	527
H 47						128	527
H 48						234	527
H 49						236	530
H 50				-9			359
H 51						67.5	540
H 52						46.0	540
H 53						50.0	540
H 54		0.86	0.91 I 0.73II	-50	-23_{170} -130_{500}		372, 419, 47, 97 419, 97, (20), (46)

No.	Polymer	$\dfrac{\delta}{\text{cal}^{1/2}\,\text{cm}^{-3/2}}$	$\dfrac{\varDelta H_u}{\text{cal mol}^{-1}}$	n
H 55	1-Hexene, 4,4-dimethyl-, (stereoregular)			
H 56	—, 5,5-dimethyl-			
H 57	—, 4-methyl-, (isotactic)			
H 58	—, 5-methyl-, (isotactic)			
H 59	Hydracrylic acid, polyester			
H 60	—, 3-(p-chlorophenyl)-2,2-dimethyl-, polyester (stereoregular)			
H 61	—, 2,2-dimethyl-3-(m-nitrophenyl)-, polyester (stereoregular)			
H 62	—, 2,2-dimethyl-3-phenyl-, polyester (stereoregular)			
H 63	—, 3-(2-furyl)-2,2-dimethyl-, polyester (stereoregular)			
H 64	—, 3-(p-methoxyphenyl)-2,2-dimethyl-, polyester (stereoregular)			
H 65	Hydroquinone, 2-hydroxytrimethylene diether with 4,4'-isopropylidenediphenol, 2-hydroxytrimethylene polyether (ar, ar', alt)			
H 66	—, 2-hydroxytrimethylene polyether (alt)			
H 67	—, —, polyacetate			
I 1	5-Indanol, 3-(p-hydroxyphenyl)-1,1,3-trimethyl-, 2-hydroxytrimethylene polyether (alt)			
I 2	Indene			
I 3	Isobutylene. See Propene, 2-methyl-.			
I 4	Isobutylene oxide. See Propane, 1,2-epoxy-2-methyl-.			
I 5	Isobutyraldehyde, polyacetal (isotactic)			
I 6	Isocyanic acid, butyl ester. See Carbamic acid, butyl-, polyamide.			
I 7	—, decamethylene ester, polyurethan with 1,10-decanediol			
I 8	—, —, polyurethan with 1,12-dodecanediol			
I 9	—, —, polyurethan with ethylene glycol			
I 10	—, —, polyurethan with 1,16-hexadecanediol			
I 11	—, ethylenebis(thioethylene) ester, polyurethan with 1,4-butanediol			
I 12	—, —, polyurethan with trans-1,4-cyclohexanedimethanol			
I 13	—, —, polyurethan with trans-1,4-cyclohexanediol			
I 14	—, —, polyurethan with ethylene glycol			
I 15	—, —, polyurethan with 1,6-hexanediol			

No.	$\dfrac{10^4 \, d\bar{v}/dT}{cc \, g^{-1} \, deg^{-1}}$	d_a $g\,cm^{-3}$	d_c $g\,cm^{-3}$	T_g °C	T_d °C	T_m °C	References
H 55						>350	117
H 56				53s			292
H 57			(0.86)			188	76, 66, (117)
H 58		(0.85)	0.84	−14s		130	76, 66, 292, (117)
H 59			$(1.425)^{21}$			122	152, (411)
H 60						260	233
H 61						240	233
H 62						290	233
H 63						180	233
H 64						240	233
H 65					$95_{0.083}$		342
H 66					$60_{0.083}$		342
H 67					$49_{0.083}$		342
I 1					$120_{0.083}$		342
I 2			(1.10)				33
I 3							
I 4							
I 5						>260	405
I 6							
I 7						145	157, (559)
I 8				34		133	559
I 9			(1.1750)	60.8		174	466
I 10				34		128	559
I 11						177	378
I 12						121	378
I 13						214	378
I 14						168	378
I 15						140	378

No.	Polymer	$\dfrac{\delta}{\text{cal}^{1/2}\,\text{cm}^{-3/2}}$	$\dfrac{\Delta H_u}{\text{cal mol}^{-1}}$	n
I 16	**Isocyanic acid,** ethylenebis (thioethylene) ester, polyurethan with 1,5-pentanediol			
I 17	−, −, polyurethan with 1,3-propanediol			
I 18	−, ethylenedi-p-phenylene ester, polyurethan with ethylene glycol			
I 19	−, hexamethylenebis(thioethylene) ester, polyurethan with 1,4-butanediol			
I 20	−, −, polyurethan with $trans$-1,4-cyclohexanedimethanol			
I 21	−, −, polyurethan with $trans$-1,4-cyclohexanediol			
I 22	−, −, polyurethan with ethylene glycol			
I 23	−, −, polyurethan with 1,6-hexanediol			
I 24	−, −, polyurethan with 1,5-pentanediol			
I 25	−, −, polyurethan with 1,3-propanediol			
I 26	−, hexamethylene ester, polyurethan with 1,4-butanediol			
I 27	−, −, polyurethan with cis-1,2-cyclopropanedimethanol			
I 28	−, −, polyurethan with $trans$-1,2-cyclopropanedimethanol			
I 29	−, −, polyurethan with 1,10-decanediol			
I 30	−, −, polyurethan with 1,12-dodecanediol			
I 31	−, −, polyurethan with 1,16-hexadecanediol			
I 32	−, −, polyurethan with 3,3,3′,3′,5,5′-hexamethyl-1,1′-spirobi[indan]-6,6′-diol			
I 33	−, −, polyurethan with 1,6-hexanediol			
I 34	−, −, polyurethan with 1,8-octanediol			
I 35	−, −, polyurethan with 3,3′-(p-phenylene)dipropanol			
I 36	−, −, polyurethan with 1,4-piperazinediethanol			
I 37	−, 4-hydroxycyclohexyl ester, polyurethan			
I 38	−, 5-hydroxypentyl ester, polyurethan			
I 39	−, methylenedi-p-phenylene ester, polyurethan with 1,4-butanediol			
I 40	−, −, polyurethan with cis-1,2-cyclopropanedimethanol			
I 41	−, −, polyurethan with $trans$-1,2-cyclopropanedimethanol			
I 42	−, methylenedi-p-phenylene ester, polyurethan with 1,10-decanediol			
I 43	−, −, polyurethan with 1,12-dodecanediol			
I 44	−, −, polyurethan with ethylene glycol			
I 45	−, −, polyurethan with 1,16-hexadecanediol			

No.	$\dfrac{10^4\ d\bar{v}/dT}{\text{cc g}^{-1}\ \text{deg}^{-1}}$	$\dfrac{d_a}{\text{g cm}^{-3}}$	$\dfrac{d_c}{\text{g cm}^{-3}}$	$\dfrac{T_g}{°C}$	$\dfrac{T_d}{°C}$	$\dfrac{T_m}{°C}$	References
I 16						130	378
I 17						157	378
I 18			(1.3055)	116.9		312	466, 333
I 19						117	378
I 20						124	378
I 21						204	378
I 22						129.5	378
I 23						106	378
I 24						94	378
I 25						115	378
I 26					40_1	180	16, 157
					-135_1		146
I 27						120	340
I 28						165	340
I 29						148	559
I 30			36			139	559
I 31			39			134	559
I 32					$240_{0.1}$		455
I 33						164	559, (157)
I 34						152	559
I 35						240	157
I 36						165	157
I 37						>335	210
I 38						155	210, (157)
I 39						194	559
I 40						230	340
I 41						290	340
I 42			48			166	559
I 43			43			164	559
I 44			(1.3233)	94.2		239	466, 333
I 45			40			152	559

No.	Polymer	$\dfrac{\delta}{\text{cal}^{1/2}\,\text{cm}^{-3/2}}$	$\dfrac{\varDelta\,H_u}{\text{cal mol}^{-1}}$	n
I 46	**Isocyanic acid,** methylenedi-p-phenylene ester, polyurethan with 1,6-hexanediol			
I 47	–, –, polyurethan with 1,8-octanediol			
I 48	–, methylphenylene ester, polyurethan with cis-1,2-cyclopropanedimethanol			
I 49	–, –, polyurethan with $trans$-1,2-cyclopropanedimethanol			
I 50	–, nonamethylene ester, polyurethan with ethylene glycol			
I 51	–, octamethylene ester, polyurethan with $trans$-1,4-cyclohexanedimethanol			
I 52	–, –, polyurethan with 1,4-cyclohexanediol			
I 53	–, –, polyurethan with 1,10-decanediol			
I 54	–, –, polyurethan with 1,6-hexanediol			
I 55	–, –, polyurethan with 2,2'-p-phenylene diethanol			
I 56	–, pentamethylenebis(thioethylene) ester, polyurethan with 1,4-butanediol			
I 57	–, –, polyurethan with $trans$-1,4-cyclohexanedimethanol			
I 58	–, –, polyurethan with $trans$-1,4-cyclohexanediol			
I 59	–, –, polyurethan with ethylene glycol			
I 60	–, –, polyurethan with 1,6-hexanediol			
I 61	–, –, polyurethan with 1,5-pentanediol			
I 62	–, –, polyurethan with 1,3-propanediol			
I 63	–, p-phenylenedimethylene ester, polyurethan with 1,4-butanediol			
I 64	–, –, polyurethan with 1,10-decanediol			
I 65	–, –, polyurethan with 1,12-dodecanediol			
I 66	–, –, polyurethan with 1,16-hexadecanediol			
I 67	–, –, polyurethan with 1,6-hexanediol			
I 68	–, –, polyurethan with 1,8-octanediol			
I 69	–, phenyl ester. See Carbanilic acid, polyamide.			
I 70	–, tetramethylenebis(thioethylene)ester, polyurethan with 1,4-butanediol			
I 71	–, –, polyurethan with $trans$-1,4-cyclohexanedimethanol			
I 72	–, –, polyurethan with $trans$-1,4-cyclohexanediol			
I 73	–, –, polyurethan with ethylene glycol			
I 74	–, –, polyurethan with 1,6-hexanediol			
I 75	–, –, polyurethan with 1,5-pentanediol			
I 76	–, –, polyurethan with 1,3-propanediol			

No.	$\dfrac{10^4\,d\bar{v}/dT}{\text{cc g}^{-1}\,\text{deg}^{-1}}$	$\dfrac{d_a}{\text{g cm}^{-3}}$	$\dfrac{d_c}{\text{g cm}^{-3}}$	$\dfrac{T_g}{°C}$	$\dfrac{T_d}{°C}$	$\dfrac{T_m}{°C}$	References
I 46				51		179	559
I 47						172	559
I 48						170	340
I 49						200	340
I 50			(1.1750)	44.0		168	466
I 51						>160	157
I 52						221	157
I 53						138	157
I 54						153	157
I 55						212	157
I 56						113	378
I 57						101.5	378
I 58						189	378
I 59						126.5	378
I 60						98	378
I 61						87	378
I 62						112	378
I 63						227	559
I 64				49		184	559
I 65				45		178	559
I 66				47		168	559
I 67				56		209	559
I 68						196	559
I 69							
I 70						119	378
I 71						129	378
I 72						214	378
I 73						126.5	378
I 74						113	378
I 75						82	378
I 76						113	378

No.	Polymer	$\dfrac{\delta}{\text{cal}^{1/2}\,\text{cm}^{-3/2}}$	$\dfrac{\Delta H_u}{\text{cal mol}^{-1}}$	n
I 77	**Isocyanic acid,** tetramethylenedi-p-phenylene ester, polyurethan with ethylene glycol			
I 78	−, tetramethylene ester, polyurethan with 1,4-butanediol			
I 79	−, trimethylenebis(thioethylene)ester, polyurethan with $trans$-1,4-cyclohexanedimethanol			
I 80	−, −, polyurethan with $trans$-1,4-cyclohexanediol			
I 81	−, trimethylenedi-p-phenylene ester, polyurethan with ethylene glycol			
I 82	**Isophthalic acid,** polyamide with $trans$-1,4-cyclohexanebis(methylamine)			
I 83	−, polyamide with fluorene-9,9-bis(propylamine)			
I 84	−, polyamide with 1,6-hexanediamine			
I 85	−, polyamide with N,N'-m-phenylenebis(m-aminobenzamide)			
I 86	−, polyamide with N,N'-m-phenylenebis(p-aminobenzamide)			
I 87	−, polyamide with N,N'-p-phenylenebis(m-aminobenzamide)			
I 88	−, polyamide with N,N'-p-phenylenebis(p-aminobenzamide)			
I 89	−, polyamide with m-phenylenediamine			
I 90	−, polyamide with p-phenylenediamine			
I 91	−, polyamide with piperazine			
I 92	−, polyamide with m-xylene-α,α'-diamine			
I 93	−, polyamide with p-xylene-α,α'-diamine			
I 94	−, polyanhydride			
I 95	−, polyester with 4,4'-benzylidenediphenol			
I 96	−, polyester with 3,3-bis(p-hydroxyphenyl)-2-indolinone			
I 97	−, polyester with 3,3-bis(p-hydroxyphenyl)-2-methylphthalimidine			
I 98	−, polyester with 3,3-bis(p-hydroxyphenyl)phthalimidine			
I 99	−, polyester with 1,4-butanediol		10100	
I 100	−, polyester with 4,4'-sec-butylidenediphenol			
I 101	−, polyester with $trans$-1,4-cyclohexanedimethanol			
I 102	−, polyester with 4,4'-cyclohexylidenediphenol			
I 103	−, polyester with $trans$-1,2-cyclopropanedimethanol			
I 104	−, polyester with 5,7-dichloro-3,3-bis(p-hydroxyphenyl)-2-indolinone			
I 105	−, polyester with 4,4'-(1,2-dimethylpropylidene)diphenol			
I 106	−, polyester with 4,4'-(diphenylmethylene)diphenol			

No.	$\dfrac{10^4\ d\bar{v}/dT}{\text{cc g}^{-1}\text{ deg}^{-1}}$	$\dfrac{d_a}{\text{g cm}^{-3}}$	$\dfrac{d_c}{\text{g cm}^{-3}}$	$\dfrac{T_g}{°C}$	$\dfrac{T_d}{°C}$	$\dfrac{T_m}{°C}$	References
I 77				105.7	274	466	
I 78					193	157	
I 79					105	378	
I 80					202	378	
I 81				74.1	207	466	
I 82					310	410	
I 83				175s		432	
I 84				130s		432	
I 85		(1.35)		290	410	507	
I 86		(1.36)		335	475	507	
I 87				305	460	507	
I 88					480	507	
I 89					390	249	
I 90					500	249	
I 91				192s	265s	432, 85	
I 92				165s		432	
I 93					290	410	
I 94					130_{100}	179	79, 229
I 95				170s		154	
I 96				256		362	
I 97				285		362	
I 98				325		362	
I 99	1.268^{25}		$(1.309)^{25}$		152.5	251	
	1.14^{152}					251	
I 100				128		115, (154)	
I 101					197	384	
I 102				127		362, (115)	
I 103					100	340	
I 104				270		362	
I 105				145s		154	
I 106				200s		154	

No.	Polymer	$\dfrac{\delta}{\mathrm{cal}^{1/2}\,\mathrm{cm}^{-3/2}}$	$\dfrac{\Delta H_u}{\mathrm{cal\ mol}^{-1}}$	n
I 107	**Isophthalic acid,** polyester with ethylene glycol			
I 108	−, polyester with 4,4'-(1-ethylpropylidene)diphenol			
I 109	−, polyester with fluorescein			
I 110	−, polyester with 1,6-hexanediol			
I 111	−, polyester with 4,4'-isopropylidenediphenol			
I 112	−, polyester with 4,4'-(α-methylbenzylidene)diphenol			
I 113	−, polyester with phenolphthalein			
I 114	−, polyester with 1,3-propanediol			
I 115	−, polyester with 3,3,3′,3′-tetramethyl-1,1′-spirobi[indan]-6,6′-diol			
I 116	−, polyester with 4,4'-(2,2,2-trichloroethylidene)diphenol			
I 117	−, polyester with m-xylene-α,α'-diol			
I 118	−, polyester with p-xylene-α,α'-diol			
I 119	−, **5-$tert$-butyl-,** polyester with phenolphthalein			
I 120	−, **5-chloro-,** polyester with phenolphthalein			
I 121	**Isoprene**			$1.5205_D^{20.5}$
I 122	−, cis-1,4-polymer	7.9	1040	1.520^{20}
		7.98	1050	1.51909_D^{25}
		8.15		
		8.35		
I 123	−, $trans$-1,4-polymer		$3070\,\alpha$	
			$< 2000\,\beta$	
I 124	**Isovaleraldehyde,** polyacetal			

No.	$10^4 \, d\bar{v}/dT$ cc g^{-1} deg^{-1}	d_a g cm^{-3}	d_c g cm^{-3}	T_g °C	T_d °C	T_m °C	References
I 107	(0.95)/(3.85)	1.346^{25}	(1.38)	51		143	32, 251, 414
	(2.00)/(5.32)	1.338					102, 131, (180)
		1.335					414
I 108				130s			154
I 109				276			362
I 110						140	251
I 111				180		285	362, 115, (154)
I 112				190s			154
I 113				318			362
I 114						132	251
I 115					260$_{0.1}$		455
I 116				150s			154
I 117				143s			432
I 118						100	384
I 119				279			362
I 120				313			362
I 121				−48			129, 112
I 122	2.07/6.16	0.909^{28}	1.00^{25}	−67	−50$_{1.2}$	36	11, 58, 326, 2, 38, 117
	2.0/6.0	0.910^{25}	0.999^{28}		−135$_{11}$		16, 66, 13, 572, 186
	/7.42	0.9060^{20}	1.007				7, 192, 283
							9
I 123	0.75/8.2	0.904^{20}	1.05$^{20}\alpha$	−68		74α	59, 335, 542, 369
	/8.33	0.870^{80}	1.04$^{20}\beta$			64β	(322)
		0.862^{85}	1.045β			48.5γ	283, 542
I 124			1.037				195

No.	Polymer	$\dfrac{\delta}{\text{cal}^{1/2}\,\text{cm}^{-3/2}}$	$\dfrac{\varDelta H_u}{\text{cal mol}^{-1}}$	n
K 1	**Ketene, dimethyl,-** polyester. See 3-**Pentenoic acid,** 3-**hydroxy-2,2,4-trimethyl-**, polyester.			
K 2	−, −, polyketone			
K 3	**Ketone,** *tert*-**butyl vinyl.** See 1-**Penten-3-one, 4,4-dimethyl-**.			
K 4	−, **cyclohexyl vinyl,** (stereoregular)			
K 5	−, **ethyl vinyl.** See 1-**Penten-3-one.**			
K 6	−, **isopropenyl methyl.** See 3-**Buten-2-one, 3-methyl-**.			
K 7	−, **isopropyl vinyl.** See 1-**Penten-3-one, 4-methyl-**.			
K 8	−, **methyl vinyl.** See 3-**Buten-2-one.**			
K 9	−, **phenyl vinyl.** See **Acrylophenone.**			
L 1	**Lauric acid,** vinyl ester			
M 1	**Malonic acid,** polyester with diethylene glycol			
M 2	−, polyester with 2,2-dimethyl-1,3-propanediol			
M 3	−, polyester with 1,3-propanediol			
M 4	−, **heptyl-**, polyester with diethylene glycol			
M 5	−, **methyl-**, polyester with diethylene glycol			
M 6	−, **methylene-**, dimethyl ester			
M 7	−, **nonyl-**, polyester with diethylene glycol			
M 8	−, **pentyl-**, polyester with diethylene glycol			
M 9	−, **propyl-**, polyester with diethylene glycol			
M 10	**Methacrylamide,** *N*-**benzyl-**			1.5965_D^{20}
M 11	−, *N*-**butyl-**			1.5135_D^{20}
M 12	−, *N*-*tert*-**butyl-**			
M 13	−, *N*-(2-**ethylhexyl**)-			
M 14	−, *N*-(2-**methoxyethyl**)-			1.5246_D^{20}
M 15	−, *N*-**methyl-**			1.5398_D^{20}
M 16	−, *N*-(2-**phenethyl**)-			1.5857_D^{20}

No.	$\dfrac{10^4\,d\bar{v}/dT}{cc\ g^{-1}\ deg^{-1}}$	$\dfrac{d_a}{g\ cm^{-3}}$	$\dfrac{d_c}{g\ cm^{-3}}$	$\dfrac{T_g}{°C}$	$\dfrac{T_d}{°C}$	$\dfrac{T_m}{°C}$	References
K 1							
K 2						250	107
K 3							
K 4						260	201
K 5							
K 6							
K 7							
K 8							
K 9							
L 1						1	353
M 1				−29			51
M 2					67		358
M 3					33		358
M 4				−58			51
M 5				−29			51
M 6	0.8/3.3			66			460
M 7				−59			51
M 8				−47			51
M 9				−38			51
M 10							311
M 11				155s			311, 199
M 12				160			93
M 13				80s			199
M 14							311
M 15							311
M 16							311

7*

No.	Polymer	δ $cal^{1/2}\,cm^{-3/2}$	ΔH_u $cal\,mol^{-1}$	n
M 17	**Methacrylic acid,** 2-aminoethyl ester			1.537_D^{20}
M 18	−, benzyl ester	9.9		1.5680_D^{20}
				1.5678
M 19	−, bornyl ester			1.5059_D^{20}
M 20	−, 2-bromoethyl ester			1.5426_D^{20}
M 21	−, p-bromophenyl ester			1.5964_D^{20}
M 22	−, butyl ester	8.7		1.483_D^{20}
		8.75		1.4831_D
		9.00		
M 23	−, −, (isotactic)			
M 24	−, sec-butyl ester			
M 25	−, $tert$-butyl ester	8.3		1.4638_D^{20}
M 26	−, 2-($tert$-butylamino)ethyl ester			
M 27	−, o-chlorobenzyl ester			1.5823_D^{20}
M 28	−, 2-chloro-1-(chloromethyl)ethyl ester			1.5270_D^{20}
M 29	−, 2-chlorocyclohexyl ester			1.5179_D^{20}
				1.5199_D^{20}
M 30	−, 4-chlorocyclohexyl ester			
M 31	−, o-chlorodiphenylmethyl ester			1.6040_D^{20}
M 32	−, 2-chloroethyl ester			1.517_D^{20}
M 33	−, 1-(o-chlorophenyl)ethyl ester			1.5624_D^{20}
M 34	−, 2-cyanoethyl ester			
M 35	−, p-cyanophenyl ester			
M 36	−, α-cyano-p-tolyl ester			
M 37	−, cycloheptyl ester			
M 38	−, cyclohexyl ester			1.50645_D^{20}
				1.5066_D^{20}
M 39	−, −, (isotactic)			

100

No.	$10^4\, d\bar{v}/dT$ $\text{cc g}^{-1}\text{deg}^{-1}$	d_a g cm^{-3}	d_c g cm^{-3}	T_g °C	T_d °C	T_m °C	References
M 17							311
M 18	1.7/5.0	1.179		54			234, 311, 460
							199
M 19							311
M 20	0.99/3.2			52			199, 460, 235
M 21							311
M 22	3.80/6.10	1.058^{20}		27	-150_{200}		234, 311, 31, 176, 82, 175
	3.7/6.3	1.053^{25}					328, 199, 2, (20), (318)
		1.004^{100}					477, (235), (62), (17)
M 23				-24			318
M 24	3.5/6.6	1.052^{20}		60	-140_{200}		460, 176, 235, 175
M 25	2.8/7.2	1.022^{20}		107			234, 311, 460, 176, 235
		1.033^{21}					311, (31)
M 26	2.0/5.7			33			460, 235
M 27				80s			311
M 28							311
M 29				65s	160_{300}		311, 257
					-20_{300}		311, 257
M 30					155_{300}		257
					-20_{300}		257
M 31							311
M 32		1.32		118s	-120_{200}		311, 175
M 33		1.269^{20}					311
M 34	1.0/3.1			91			460
M 35	1.1/4.1			155			460
M 36	1.8/4.8			128			460
M 37					-153_{6860}		547
M 38	2.7/	1.100		104	100_2		199, 311, 175, 93, 256
		1.0951			-20_{200}		(235), (318)
					-80_1		256
M 39				51			318

No.	Polymer	$\dfrac{\delta}{\text{cal}^{1/2}\,\text{cm}^{-3/2}}$	$\dfrac{\Delta H_u}{\text{cal mol}^{-1}}$	n
M 40	**Methacrylic acid,** 2-cyclohexylcyclohexyl ester			1.5250_D^{20}
				1.5191_D^{20}
				1.5219_D
M 41	—, p-cyclohexylphenyl ester			1.5575_D^{20}
M 42	—, cyclooctyl ester			
M 43	—, decahydro-2-naphthyl ester			
M 44	—, decyl ester			
M 45	—, 2,3-dibromopropyl ester			1.5739_D^{20}
M 46	—, 2-(diethylamino)ethyl ester			1.5174_D^{20}
M 47	—, 1,1-diethylpropyl ester			1.4889_D^{20}
M 48	—, 2-(dimethylamino)ethyl ester			
M 49	—, 1,3-dimethylbutyl ester			
M 50	—, 3,3-dimethylbutyl ester			
M 51	—, 2,2-dimethyl-2-(5,5-dimethyl-m-dioxan-2-yl)ethyl ester			
M 52	—, 1,2-diphenylethyl ester			1.5816_D^{20}
M 53	—, diphenylmethyl ester			1.5933_D^{20}
M 54	—, docosyl ester			
M 55	—, dodecyl ester	8.2		1.4740_D^{30}
M 56	—, ester with ethyl glycolate			1.4903_D^{20}
M 57	—, ester with 2-hydroxytrimethylene diacetate			1.4855_D^{20}
M 58	—, ester with methyl salicylate			1.5707_D^{20}
M 59	—, ester with phenyl salicylate			1.6006_D^{20}
M 60	—, 2-ethoxyethyl ester	9.0		
M 61	—, ethyl ester	8.95		1.485_D^{20}
		9.15		
M 62	—, —, (isotactic)			
M 63	—, 2-ethylbutyl ester			
M 64	—, 2-ethylhexyl ester			

No.	$10^4\ d\bar{v}/dT$ $\mathrm{cc\ g^{-1}\ deg^{-1}}$	d_a $\mathrm{g\ cm^{-3}}$	d_c $\mathrm{g\ cm^{-3}}$	T_g °C	T_d °C	T_m °C	References
M 40		1.075					311
							311
							199
M 41		1.115		145s			311, 199
M 42					-53_{5800}		547
M 43				75s			199
M 44				<−55			62, (17)
M 45							311
M 46							311
M 47							311
M 48	2.7/6.0			19			460, 235
	3.7/5.8						460
M 49		1.005^{20}		59s			176, 175
M 50		1.001^{20}		59s	-130_{200}		176, 175
M 51				104			493
M 52		1.147					311
M 53		1.168					311
M 54		0.857^{0}				55	255
M 55	3.8/6.80	0.929^{25}		−55			234, 403, 31, (17), (235)
M 56							311
M 57							311
M 58							311
M 59							311
M 60							328
M 61	2.75/5.40	1.119^{20}		66	83_{60}		328, 311, 31, 176, 318, 482
		1.124^{25}			-220_{9800}		477, 407, (175), (235)
							(117), (17), (82), (62)
M 62				12			318
M 63	/5.76	1.040^{25}		11			82
		0.994^{100}					82
M 64				−10b			17

No.	Polymer	$\dfrac{\delta}{\text{cal}^{1/2}\,\text{cm}^{-3/2}}$	$\dfrac{\Delta H_u}{\text{cal mol}^{-1}}$	n
M 65	**Methacrylic acid,** 2-(ethylsulfinyl)ethyl ester			
M 66	−, 9-fluorenyl ester			1.6318_D^{20}
				1.6363_D^{20}
M 67	−, 2-fluoroethyl ester			1.4768_D
M 68	−, furfuryl ester			1.5381_D^{20}
M 69	−, hexadecyl ester			1.4750_D^{30}
M 70	−, hexyl ester	8.6		1.4813_D^{20}
M 71	−, 2-hydroxyethyl ester			1.5119_D^{20}
M 72	−, −, benzoate			1.555_D^{20}
M 73	−, −, isobutyrate			
M 74	−, isobornyl ester	8.1		
M 75	−, −, (isotactic)			
M 76	−, −, (syndiotactic)			
M 77	−, isobutyl ester	8.65		1.477_D^{20}
M 78	−, isobutyl ester, (isotactic)			
M 79	−, isopentyl ester			
M 80	−, isopropyl ester			1.552_D
M 81	−, −, (isotactic)			
M 82	−, −, (syndiotactic)			
M 83	−, lead salt			1.645_D^{20}
M 84	−, menthyl ester			1.4890_D^{20}
M 85	−, p-methoxybenzol ester			1.552_D^{20}
M 86	−, p-methoxycarbonylphenyl ester			
M 87	−, methyl ester	9.08		1.490_D^{20}
		9.25		
		9.3		
		9.40		
		9.45		
		9.5		

* Major (glass) transition

No.	$\dfrac{10^4\,d\bar{v}/dT}{\text{cc g}^{-1}\text{ deg}^{-1}}$	$\dfrac{d_a}{\text{g cm}^{-3}}$	$\dfrac{d_c}{\text{g cm}^{-3}}$	$\dfrac{T_g}{\text{°C}}$	$\dfrac{T_d}{\text{°C}}$	$\dfrac{T_m}{\text{°C}}$	References
M 65	1.5/4.6			25			460
M 66				85s			311
							311
M 67							199
M 68				60s			311
M 69						22	403, (17)
M 70	4.6/6.6	1.007^{25}		−5			234, 311, 31, 235
		0.959^{100}					82
M 71	1.02/2.6			55			311, 460, 235
M 72							311
M 73	3.27/5.74			2			188
M 74				110			234, 235
M 75				110			318
M 76				111			318
M 77	2.6/6.0	1.045^{20}		53			477, 311, 460, 176, 318
	2.3/6.4						460, (17), (235)
M 78				8			318
M 79		1.032^{20}		46s			176
M 80	2.1/6.7	1.033^{20}		81			199, 460, 176, 318, 235, (175)
M 81				27			318
M 82				85			318
M 83							311
M 84		1.010		90s			311
M 85							311
M 86	1.7/4.6			106			460
M 87	2.3/5.2	1.195^{0}		114*	$120_{0.12}$		2, 311, 514, 32, 516, 16
	1.7/2.3	1.17^{25}		60	36_{10}		7, 31, 513, 16
	1.2/1.7	1.188^{25}		−7	-82_1		234, 144, (87)
		1.196^{25}					477, 50, (175), (235)
							328, (310), (498)
							182

No.	Polymer	$\dfrac{\delta}{\text{cal}^{1/2}\,\text{cm}^{-3/2}}$	$\dfrac{\Delta H_u}{\text{cal mol}^{-1}}$	n
M 88	**Methacrylic acid,** methyl ester, (isotactic)			
M 89	−, −, (syndiotactic)			
M 90	−, 1-methylbutyl ester			
M 91	−, 1-methylcyclohexyl ester			1.511_D^{20}
				1.510_D
M 92	−, 2-methylcyclohexyl ester			1.5028_D^{20}
M 93	−, 3-methylcyclohexyl ester			1.4947_D^{20}
M 94	−, 4-methylcyclohexyl ester			1.4975_D^{20}
M 95	−, (5-methyl-m-dioxan-5-yl)methyl ester			
M 96	−, 1-methylheptyl ester			
M 97	−, 2-methyl-2-nitropropyl ester			1.4868_D^{20}
M 98	−, 1-methylpentyl ester			
M 99	−, 1-naphthyl ester			1.6410_D^{20}
				1.6395_D^{20}
M 100	−, 2-naphthyl ester			1.6298_D^{20}
M 101	−, 1-naphthylmethyl ester			1.6316_D^{20}
M 102	−, neopentyl ester			
M 103	−, m-nitrobenzyl ester			1.5845_D^{20}
M 104	−, octadecyl ester	7.8		
M 105	−, octyl ester	8.4		
M 106	−, pentabromophenyl ester			1.71_D^{20}
M 107	−, pentachlorophenyl ester			1.608_D^{20}
M 108	−, pentyl ester			
M 109	−, 2-phenoxyethyl ester			1.5570_D^{20}
M 110	−, phenyl ester			1.7515_D^{20}
M 111	−, 1-phenylcyclohexyl ester			1.5645_D^{20}
M 112	−, 1-phenylethyl ester			1.5487_D^{20}
M 113	−, 2-phenethyl ester			1.5592_D^{20}

106

No.	$\dfrac{10^4\,d\bar{v}/dT}{\text{cc g}^{-1}\text{ deg}^{-1}}$	$\dfrac{d_a}{\text{g cm}^{-3}}$	$\dfrac{d_c}{\text{g cm}^{-3}}$	$\dfrac{T_g}{°\text{C}}$	$\dfrac{T_d}{°\text{C}}$	$\dfrac{T_m}{°\text{C}}$	References
M 88		1.22^{30}	1.23	48	$-95_{1.1}$	160	87, 517, 516
		$(1.2063)^{25}$	$(1.238)^{25}$				108, 144
M 89		1.19^{30}		126	$132_{0.90}$	200	87, 516, 144, 76
					$-95_{1.2}$		144
M 90		1.030^{20}		63s			176
M 91							311
							199
M 92				70s			311
M 93				70s			311
M 94				68s	-20_{1000}		311, 257
M 95				142			493
M 96		0.988^{20}					176
M 97				150s			311, 199
M 98		1.013^{20}		23s			176
M 99				75s			311
							311
M 100							311
M 101				70s			311
M 102		0.993^{20}		115s			176
M 103		1.339					311
M 104			(0.97)		-50_{200}	37.5	234, 175, 31, 20
					-120_{200}		175
M 105	4.4/5.8	0.971^{25}		-20			234, 31, 235, (17), (62)
M 106							311
M 107				150s			311
M 108		1.032^{20}		-5b			176, 17
M 109							311
M 110	1.6/5.3	1.21		120			311, 460, 175, 93, (235)
M 111							311
M 112		1.129^{20}					311
M 113	1.8/5.1			26			311, 460

No.	Polymer	$\dfrac{\delta}{\text{cal}^{1/2}\,\text{cm}^{-3/2}}$	$\dfrac{\Delta H_u}{\text{cal mol}^{-1}}$	n
M 114	**Methacrylic acid,** 1-phenylpentyl ester			1.5396_D^{20}
M 115	−, 2-(phenylsulfonyl)ethyl ester			1.5682_D^{20}
M 116	−, propargyl ester. See Methacrylic acid, 2-propynyl ester			
M 117	−, propyl ester	8.8		1.484_D^{20}
M 118	−, 2-propynyl ester			
M 119	−, tetradecyl ester			1.4746_D^{30}
M 120	−, tetrahydrofurfuryl ester			1.5096_D^{20}
				1.51081_D^{20}
M 121	−, m-tolyl ester			1.5683_D^{20}
M 122	−, o-tolyl ester			1.5707_D^{20}
M 123	−, 2,2,2-trifluoroethyl ester			1.437_D
M 124	−, 2,2,2-trifluoro-1-methylethyl ester			1.4185_D
M 125	−, 3,3,5-trimethylcyclohexyl ester			1.485_D^{20}
M 126	−, (2,2,5-trimethyl-m-dioxan-5-yl)methyl ester			
M 127	−, 1,2,2-trimethylpropyl ester			
M 128	−, **thio-,** S-butyl ester			1.5390_D^{20}
M 129	**Methacrylonitrile**	10.7		
M 130	**Methane,** 4,4′-**bis**(p-carboxyphenyl)-. See **Benzoic acid,** 4,4′-methylenedi-.			
M 131	−, 4,4′-**bis**(p-carboxyphenoxy)-. See **Benzoic acid,** 4,4′-methylenedioxydi-.			
M 132	−, 4,4′-**bis**(p-hydroxyphenyl)-. See **Phenol,** 4,4′-methylenedi-.			
M 133	**Methylene.** See **Ethylene.**			
M 134	**Muconic acid.** See **2,4-Hexadienedioic acid.**			
M 135	**Myristic acid,** vinyl ester			

No.	$\dfrac{10^4\ d\bar{v}/dT}{cc\ g^{-1}\ deg^{-1}}$	$\dfrac{d_a}{g\ cm^{-3}}$	$\dfrac{d_c}{g\ cm^{-3}}$	$\dfrac{T_g}{°C}$	$\dfrac{T_d}{°C}$	$\dfrac{T_m}{°C}$	References
M 114							311
M 115				45s			311
M 116							
M 117	3.15/5.8	1.085^{20}		35	-150_{200}		234, 311, 31, 176, 175
		1.077^{25}					31, (17), (62), (20)
M 118		1.17					175
M 119				-72		-9	403, 62, (17)
M 120				60s			311
							311
M 121							311
M 122				106s			311, 199
M 123							199
M 124		1.34		81s			199, 175
M 125				79s			311
M 126				100			493
M 127		0.991^{20}		123s			176
M 128							311
M 129		1.10		120			7, 2, 199, 3, (117)
M 130							
M 131							
M 132							
M 133							
M 134							
M 135						28.5	353

No.	Polymer	$\dfrac{\delta}{\text{cal}^{1/2}\,\text{cm}^{-3/2}}$	$\dfrac{\Delta H_u}{\text{cal mol}^{-1}}$	n
N 1	**Naphthalene, 4-chloro-1-vinyl-**			
N 2	−, **1-vinyl-**			1.6818_D^{20}
N 3	−, −, (isotactic)			
N 4	**1,4-Naphthalenedicarboxylic acid,** polyester with ethylene glycol			
N 5	**1,5-Naphthalenedicarboxylic acid,** polyester with ethylene glycol			
N 6	**2,6-Naphthalenedicarboxylic acid,** polyester with *cis*-1,4-cyclohexanedimethanol			
N 7	−, polyester with ethylene glycol			
N 8	−, polyester with *p*-xylene-α, α'-diol			
N 9	**2,7-Naphthalenedicarboxylic acid,** polyester with ethylene glycol			
N 10	**Nonanedioic acid.** See **Azelaic acid.**			
N 11	**Nonanoic acid, 9-amino-,** polyamide			
N 12	−, **9-(4-carboxybutoxy)-,** polyamide with 1,6-hexanediamine			
N 13	−, −, polyamide with 5,5′oxybis(pentylamine)			
N 14	−, −, polyamide with *p*-xylene-α, α'-diamine			
N 15	−, **9-(3-carboxypropoxy)-,** polyamide with 1,6-hexanediamine			
N 16	−, −, polyamide with 5,5′-oxybis(pentylamine)			
N 17	−, −, polyamide with *p*-xylene-α, α'-diamine			
N 18	−, **9,9′-oxydi-,** polyamide with 1,6-hexanediamine			
N 19	−, −, polyamide with 5,5′-oxybis(pentylamine)			
N 20	−, −, polyamide with *p*-xylene-α, α'-diamine			
N 21	**1-Nonene,** (stereoregular)			
N 22	**2,5-Norbornadiene-2-carboxylic acid,** ethyl ester			
N 14	**Norbornene.** See **Cyclopentane,** 1,3-*trans*-vinylene polymer *(alt)*			
N 24	**Nylon.** See under name of acid source material.			

No.	$\dfrac{10^4 \, d\bar{v}/dT}{\text{cc g}^{-1}\text{ deg}^{-1}}$	$\dfrac{d_a}{\text{g cm}^{-3}}$	$\dfrac{d_c}{\text{g cm}^{-3}}$	$\dfrac{T_g}{\text{°C}}$	$\dfrac{T_d}{\text{°C}}$	$\dfrac{T_m}{\text{°C}}$	References
N 1				172_s			199
N 2				162_s			311, 96
N 3			1.12			360	119, 446, 76
N 4	(1.28)/(3.62)		$(1.36)^0$	64			32
N 5	1.56/3.43	1.37^0		71		230	32, 157, 34
N 6						287	384
N 7	1.41/4.86	1.33^0		113		260	32, 157, 34
N 8						280	384
N 9	(1.12)/(4.99)		$(1.35)^0$	119		270	32, 180
N 10							
N 11	(3.6)/(4.4)	$(1.052)^{52}$	$(1.066)^{30}$	51	90_{61}	209	106, 162, (321), (348)
					-47_{134}		162, (350)
N 12					154		527
N 13					117		527
N 14					217		527
N 15					155		527
N 16					118		527
N 17					215		527
N 18					158		527
N 19					129		527
N 20					215		527
N 21					-47_{170}	19	97
					-160_{310}		97
N 22				220			98
N 23							
N 24							

No.	Polymer	$\dfrac{\delta}{\text{cal}^{1/2}\,\text{cm}^{-3/2}}$	$\dfrac{\varDelta H_u}{\text{cal mol}^{-1}}$	n
O 1	**Octadecanedioic acid,** polyamide with 1,10-decanediamine			
O 2	−, polyamide with N,N′-diethyl-p-xylene-α,α'-diamine			
O 3	−, polyamide with N,N′-dimethyl-1,6-hexanediamine			
O 4	−, polyamide with N,N′-dimethyl-p-xylene-α,α'-diamine			
O 5	−, polyamide with 1,12-dodecanediamine			
O 6	−, polyamide with 1,6-hexanediamine			
O 7	−, polyamide with 1,18-octadecanediamine			
O 8	−, polyamide with 1,8-octanediamine			
O 9	−, polyamide with 1,5-pentanediamine			
O 10	−, polyamide with 2,2′-p-phenylenebis(ethylamine)			
O 11	−, polyamide with 1,14-tetradecanediamine			
O 12	−, polyamide with p-xylene-α,α'-diamine			
O 13	−, polyanhydride			
O 14	−, polyester with 1,3-propanediol			
O 15	**Octadecanoic acid. See Stearic acid.**			
O 16	**1-Octadecene,** (isotactic)			
O 17	**Octanal,** polyacetal (isotactic)			
O 18	**Octanedioic acid. See Suberic acid.**			
O 19	**Octanoic acid, 8-amino-,** polyamide	12.7		
O 20	−, **pentadecafluoro-,** vinyl ester			1.3388_D^{25}
O 21	**1-Octene,** (isotactic)			
O 22	**1-Octene-2-carboxylic acid. See Acrylic acid, 2-hexyl-.**			
O 23	**7-Oxabicyclo[4·1·0]heptane,** polyether			
O 24	**Oxacyclobutane. See Oxetane.**			
O 25	**1,3,4-Oxadiazole,** 2,5-decamethylene polymer *(alt)*			
O 26	−, 2,5-nonamethylene polymer *(alt)*			

No.	$10^4\ d\bar{v}/dT$ $\mathrm{cc\ g^{-1}\ deg^{-1}}$	d_a $\mathrm{g\ cm^{-3}}$	d_c $\mathrm{g\ cm^{-3}}$	T_g °C	T_d °C	T_m °C	References
O 1			$(1.024)^{25}$			170	529
O 2						50	558
O 3						52	558
O 4						80	558
O 5			$(1.016)^{25}$			167	529
O 6			$(1.040)^{25}$			192	529
O 7			$(0.998)^{25}$			158	529
O 8			$(1.030)^{25}$			179	529
O 9						167	352
O 10			$(1.055)^{25}$		75_{110}	248	529
O 11			$(1.013)^{25}$			158	529
O 12			$(1.075)^{25}$		75_{110}	242	529
O 13						95	350
O 14						76	180
O 15							
O 16		$(0.883)^{25}$	0.95		55_{25} -110_{170}	75	536, 372, 419, 97 97, (117), (46)
O 17						35	405
O 18							
O 19	(3.1)/(3.5)	$(1.069)^{25}$	1.13α 1.09γ	51	49_1 $-50_{0.1}$ -130_{10}	199	182, 106, 162, 299, 16 299, 16, (321), (348) 16, (350), (354)
O 20					-10		359
O 21				-45	-42_{140} -160_{300}		76, 97, (46) 97
O 22							
O 23				70s			78
O 24							
O 25			(1.22)			100I 43II	454 454
O 26			(1.28)			82 I 45II	454 454

No.	Polymer	$\dfrac{\delta}{\text{cal}^{1/2}\,\text{cm}^{-3/2}}$	$\dfrac{\varDelta H_u}{\text{cal mol}^{-1}}$	n
O 27	**1,3,4,-Oxadiazole,** 2,5-octamethylene polymer *(alt)*			
O 28	**Oxalic acid,** polyamide with 1,10-decanediamine			
O 29	−, polyamide with 3,3′-diamino-*N*-methyldipropylamine			
O 30	−, polyamide with 3,3′-(octylphosphinidene)bispropylamine			
O 31	−, polyamide with 3,3′-(phenylphosphinidene)bispropylamine			
O 32	−, polyamide with 3,3′-(tetramethylenedioxy)bispropylamine			
O 33	−, polyester with 2,2-bis(chloromethyl)-1,3-propanediol			
O 34	−, polyester with *trans*-1,4-cyclohexanedimethanol			
O 35	−, polyester with diethylene glycol			
O 36	−, polyester with 2,2-dimethyl-1,3-propanediol			
O 37	−, polyester with ethylene glycol			
O 38	−, polyester with 1,2-propanediol			
O 39	−, polyester with 1,3-propanediol			
O 40	−, polyester with *p*-xylene-α, α′-diol			
O 41	**Oxetane,** polyether			
O 42	−, **3,3-bis(bromomethyl)-**, polyether			
O 43	−, **3,3-bis(chloromethyl)-**, polyether		5740	
			5490	
O 44	−, **3,3-bis(ethoxymethyl)-**, polyether			
O 45	−, **3,3-bis(fluoromethyl)-**, polyether			
O 46	−, **3,3-bis(hydroxymethyl)-**, polyether			
O 47	−, **3,3-bis(iodomethyl)-**, polyether			
O 48	−, **3,3-dimethyl-**, polyether			
O 49	−, **2-methyl-**, polyether			
O 50	−, **3-methyl-**, polyether			

No.	$10^4\ d\bar{v}/dT$ cc g^{-1} deg^{-1}	d_a g cm^{-3}	d_c g cm^{-3}	T_g °C	T_d °C	T_m °C	References
O 27			(1.21)			110 I	454
						52 II	454
O 28						229	157
O 29						202	157
O 30		1.086					153
O 31		1.208					153
O 32						160	157
O 33						122	462
O 34						215	384
O 35				−8			51
O 36						111	358
O 37						172	55
O 38						180	358
O 39						89	358
O 40						214	384
O 41					-64_5	35	394, 195, (20)
					-128_5		394
O 42						220	195
O 43	(1.4)/3.18	1.152^{180}	$1.47^{20}\beta$	7.5	90_{54}	188α	463, 553, 248, 262
		1.386^{20}			20_{54}	130β	570, 260, (20), (152), (237)
					-40_{54}		262, (100)
O 44						83	195
O 45						135	195
O 46						280	195
O 47						290	195
O 48						47	195
O 49					-30_5		394
					-50_5		394
					-120_5		394
O 50					-18_5		394
					-55_5		394

No.	Polymer	$\dfrac{\delta}{\mathrm{cal}^{1/2}\,\mathrm{cm}^{-3/2}}$	$\dfrac{\Delta H_u}{\mathrm{cal\ mol}^{-1}}$	n
O 51	**Oxide, decamethylene,** polyether			
O 52	—, **hexamethylene,** polyether			
O 53	**Oxirane, 2,3-dimethyl-.** See **Butane, 2,3-epoxy-.**			
P 1	**Palmitic acid,** vinyl ester			
P 2	**Pelargonic acid.** See **Nonanoic acid.**			
P 3	**Pentadecanedioic acid,** polyamide with 2,2'-p-phenylenebis(ethylamine)			
P 4	—, polyamide with p-xylene-α,α'-diamine			
P 5	*trans*-**1,3-Pentadiene,** 1,2-polymer (syndiotactic)			
P 6	**1,3-Pentadiene,** *cis*-1,4-polymer (isotactic)			
P 7	—, *cis*-1,4-polymer (syndiotactic)			
P 8	—, *trans*-1,4-polymer (isotactic)			
P 9	**Pentanal.** See **Valeraldehyde.**			
P 10	**Pentanedioic acid.** See **Glutaric acid.**			
P 11	**Pentanoic acid.** See **Valeric acid.**			
P 12	**1-Pentene**			
P 13	—, (isotactic)			
P 14	—, **5-cyclohexyl-,** (stereoregular)			
P 15	—, **4,4-dimethyl-,** (stereoregular)			
P 16	—, **5-iodo-**			
P 17	—, **3-methyl-,** (stereoregular)			
P 18	—, **4-methyl-,** (isotactic)		2850	
P 19	—, **5-phenyl-,** (stereoregular)			

---

* Major (glass) transition.
** Crystalline second-order transition.

116

No.	$\dfrac{10^4\ d\bar{v}/dT}{\text{cc g}^{-1}\ \text{deg}^{-1}}$	$\dfrac{d_a}{\text{g cm}^{-3}}$	$\dfrac{d_c}{\text{g cm}^{-3}}$	$\dfrac{T_g}{°C}$	$\dfrac{T_d}{°C}$	$\dfrac{T_m}{°C}$	References
O 51						80	120
O 52						62	540, (120)
O 53							
P 1						41.2	353
P 2							
P 3			$(1.075)^{25}$			248	529
P 4			$(1.090)^{25}$		90_{110}	241	529
P 5						10	470
P 6			(0.924)			47	391, (290)
P 7			(0.915)			53	391, 289
P 8		0.89	0.98	-60		95	141, 322, 44, 222
P 9							
P 10							
P 11							
P 12				-50			76
P 13	/9.2	0.795^{120}	0.92 I	14	-3_{80}	111 I	296, 334, 97, 408
		$(0.860)^{25}$	0.90 II	-10	-130_{300}	80 II	438, (47), (292), (117)
				-52			296, (20), (46), (40)
P 14				-25s		123	292, 401
P 15			0.90	59		>350	504, 292, 117, (46)
P 16					30_{50}		486
					-98_{210}		486
P 17			(0.83)			278	76, 164
P 18	3.83/7.61	0.838^{20}	0.828^{20}	130	38_1	250	398, 65, 312, 225
		$0.839^{27.7}$	0.822	29*	-120_{9000}	125**	383, 407, 323, (419)
			0.832		-250_{25000}		(417), (292), (438), (95)
P 19		1.05		-28s		100	160, 292, 76

No.	Polymer	$\dfrac{\delta}{\text{cal}^{1/2}\,\text{cm}^{-3/2}}$	$\dfrac{\Delta H_u}{\text{cal mol}^{-1}}$	n
P 20	1-Pentene, 5,5,5-trifluoro-, (stereoregular)			
P 21	−, 4-(trifluoromethyl)-, (stereoregular)			
P 22	−, 5-(trimethylsilyl)-, (stereoregular)			
P 23	1-Pentene-2-carboxylic acid. See Acrylic acid, 2-propyl-.			
P 24	−, 4-methyl-. See Acrylic acid, 2-isobutyl-.			
P 25	3-Pentenoic acid, 3-hydroxy-2,2,4-trimethyl-, polyester			
P 26	1-Pentene-3-one, (stereoregular)			
P 27	−, 4,4-dimethyl-, (stereoregular)			
P 28	−, 4-methyl-, (stereoregular)			
P 29	Phenol, 4,4'-(1-ethyl-1,3-cyclohexylene)di-, 2-hydroxytrimethylene polyether *(alt)*			
P 30	−, −, −, polyacetate			
P 31	−, 4,4'-isobutylidenedi-, 2-hydroxytrimethylene polyether *(alt)*			
P 32	−, 4,4'-isopropylidenebis[2-chloro-, 2-hydroxytrimethylene polyether *(alt)*			
P 33	−, 4,4'-isopropylidenebis[2,6-dibromo-, 2-hydroxytrimethylene diether with 4,4'-isopropylidenediphenol, 2-hydroxytrimethylene polyether *(ar, ar', alt)*			
P 34	−, 4,4'-isopropylidenebis[2,6-dichloro-, 2-hydroxytrimethylene diether with 4,4'-isopropylidenediphenol, 2-hydroxytrimethylene polyether *(ar, ar', alt)*			
P 35	−, −, 2-hydroxytrimethylene diether with 4,4'-methylenediphenol, 2-hydroxytrimethylene polyether *(ar, ar', alt)*			
P 36	−, −, 2-hydroxytrimethylene polyether *(alt)*			
P 37	−, −, −, polyacetate			
P 38	−, 4,4'-isopropylidenedi-, 2-hydroxytrimethylene polyether *(alt)*			
P 39	−, −, −, polyacetate			
P 40	−, −, −, polybenzoate			
P 41	−, −, −, poly(chloroacetate)			
P 42	−, −, −, poly(o-chlorobenzoate)			
P 43	−, −, −, polypropionate			
P 44	−, 4,4'-*m*-menth-6,8-ylenedi-, 2-hydroxytrimethylene polyether *(alt)*			
P 45	−, 4,4'-*p*-menth-1,2-ylenedi-, 2-hydroxytrimethylene polyether *(alt)*			
P 46	−, 4,4'-*p*-menth-1,8-ylenedi-, 2-hydroxytrimethylene polyether *(alt)*			
P 47	−, 4,4'-(α-methylbenzylidene)di-, 2-hydroxytrimethylene polyether *(alt)*			

No.	$\dfrac{10^4\ d\bar{v}/dT}{\text{cc g}^{-1}\ \text{deg}^{-1}}$	$\dfrac{d_a}{\text{g cm}^{-3}}$	$\dfrac{d_c}{\text{g cm}^{-3}}$	$\dfrac{T_g}{°C}$	$\dfrac{T_d}{°C}$	$\dfrac{T_m}{°C}$	References
P 20						228	6
P 21						258	6
P 22						133	195
P 23							
P 24							
P 25						170	107
P 26						180	201
P 27				94s		150	201, 200
P 28						200	201
P 29					$140_{0.083}$		342
P 30					$107_{0.083}$		342
P 31					$95_{0.083}$		342
P 32					$85_{0.083}$		342
P 33					$117_{0.083}$		342
P 34					$100_{0.083}$		342
P 35					$80_{0.083}$		342
P 36					$115_{0.083}$		342
P 37					$100_{0.083}$		342
P 38	1.18^{23}				$100_{0.083}$		342
P 39					$60_{0.083}$		342
P 40					$65_{0.083}$		342
P 41					$65_{0.083}$		342
P 42					$66_{0.083}$		342
P 43					$60_{0.083}$		342
P 44		1.106			$150_{0.083}$		500
P 45		1.106			$175_{0.083}$		500, 342
P 46		1.129			$135_{0.083}$		500, 342
P 47					$115_{0.083}$		342

No.	Polymer	$\dfrac{\delta}{\text{cal}^{1/2}\,\text{cm}^{-3/2}}$	$\dfrac{\Delta H_u}{\text{cal mol}^{-1}}$	n
P 48	**Phenol, 4,4'-(α-methylbenzylidene)di-,** 2-hydroxytrimethylene polyether *(alt)*, polyacetate			
P 49	−, −, −, polybenzoate			
P 50	−, **4,4'-methylenedi-**, 2-hydroxytrimethylene polyether *(alt)*			
P 51	−, **4,4'-sulfonyldi-**, 2-hydroxytrimethylene polyether *(alt)*			
P 52	−, −, −, polyacetate			
P 53	−, **4,4'-(1,1,3-trimethyltrimethylene)di-**, 2-hydroxytrimethylene polyether *(alt)*			
P 54	**Phosphonitrile, bis(dimethylamino)-**, polyphosphazene			
P 55	−, **bis(ethylamino)-** polyphosphazene			
P 56	−, **bis(phenylamino)-**, polyphosphazene			
P 57	−, **bis(2,2,2-trifluoroethoxy)-**, polyphosphazene			
P 58	−, **dichloro-**, polyphosphazene			
P 59	−, **diethoxy-**, polyphosphazene			
P 60	−, **dimethoxy-**, polyphosphazene			
P 61	−, **diphenoxy-**, polyphosphazene			
P 62	−, **dipiperidino -**, polyphosphazene			
P 63	**Phthalic acid,** polyamide with piperazine			
P 64	−, polyester with ethylene glycol			
P 65	−, polyester with 4,4'-isopropylidenediphenol			
P 66	**Phthalimide,** *N*-**vinyl-**			1.6200_D^{20}
P 67	**Pimelic acid,** polyamide with 1,4-butanediamine			
P 68	−, polyamide with *cis*-1,4-cyclohexanebis(methylamine)			
P 69	−, polyamide with *trans*-1,4-cyclohexanebis(methylamine)			
P 70	−, polyamide with 1,7-heptanediamine			
P 71	−, polyamide with 1,6-hexanediamine			
P 72	−, polyamide with 1,5-pentanediamine			
P 73	−, polyamide with spiro[3.3]heptane-2,6-diamine			

No.	$10^4\,d\bar{v}/dT$ $\text{cc g}^{-1}\,\text{deg}^{-1}$	d_a g cm^{-3}	d_c g cm^{-3}	T_g °C	T_d °C	T_m °C	References
P 48					$110_{0.083}$		342
							342
P 49					$126_{0.083}$		342
P 50					$80_{0.083}$		342
P 51					$155_{0.083}$		343
P 52					$130_{0.083}$		342
P 53					$75_{0.083}$		342
P 54					-4_1		511
P 55				30			511
P 56				91			511
P 57				−66		240I	468, 510
						70II	510
P 58		1.91^{20}	2.21	−63			471, 510
			2.222				473
P 59				−84			511, (468)
P 60				−76			468, 510
P 61				−8			468, 510
P 62				19			511
P 63						260s	85
P 64	1.74/5.93	1.338		17			102, 131, (99)
P 65				134			115
P 66				224			311, 93, (199)
P 67						233	352
P 68						191	410
P 69						293	410
P 70		(1.06)	1.108γ	-39_{191}		196	277, 162, 352, (348)
P 71					$58_{3.7}$	228s	16, 348, (352)
					-50_6		16
					$-115_{7.3}$		16
P 72						183	157, 352
P 73						280	509

No.	Polymer	$\dfrac{\delta}{\text{cal}^{1/2}\,\text{cm}^{-3/2}}$	$\dfrac{\Delta H_u}{\text{cal mol}^{-1}}$	n
P 74	**Pimelic acid,** polyamide with p-xylene-α,α'-diamine			
P 75	—, polyester with *trans*-1,4-cyclohexanedimethanol			
P 76	—, polyester with diethylene glycol			
P 77	—, polyester with 1,6-hexanediol			
P 78	—, polyester with 1,3-propanediol			
P 79	—, polyester with p-xylene-α,α'-diol			
P 80	**1,4-Piperazinediacetic acid,** polyamide with 2,2'-(ethylenedioxy)bisethylamine			
P 81	—, polyamide with 1,6-hexanediamine			
P 82	**1,4-Piperazinedicarboxylic acid,** polyester with bis(2-hydroxyethyl) terephthalate			
P 83	—, polyester with *cis*-1,2-cyclopropanedimethanol			
P 84	—, polyester with *trans*-1,2-cyclopropanedimethanol			
P 85	—, polyester with ethylene glycol			
P 86	—, polyester with 4,4'-isopropylidenediphenol			
P 87	—, polyester with 3,3,3′,3′-tetramethyl-1,1′-spirobi[indan]-6,6′-diol			
P 88	**1-Piperidinecarboxylic acid, 4-hydroxy-,** polyester			
P 89	**Pivalic acid,** vinyl ester			
P 90	L-**Proline,** polyamide			
P 91	**Propanal. See Propionaldehyde.**			
P 92	**Propane, 1-(allyloxy)-2,3-epithio-,** poly(thioether)			1.553_D^{30}
P 93	—, **1-(allyloxy)-2,3-epoxy-,** polyether			
P 94	—, **3-(4-biphenylyloxy)-1,2-epoxy-,** polyether			
P 95	—, **2,2-bis(p-carboxyphenyl)-.** See **Benzoic acid, 4,4'-isopropylidenedi-.**			
P 96	—, **2,2-bis(p-hydroxyphenyl)-.** See **Phenol, 4,4'-isopropylidenedi-.**			
P 97	—, **1-chloro-2-chloromethyl-2,3-epoxy-,** polyether			
P 98	—, **1-chloro-2,3-epoxy-2-methyl-,** polyether			
P 99	—, **3-(o-chlorophenoxy)-1,2-epoxy-,** polyether			
P 100	—, **3-(p-chlorophenoxy)-1,2-epoxy-,** polyether			
P 101	—, **1,2-epithio-.** See **Propylene sulfide.**			
P 102	—, **1,2-epoxy-.** See **Propylene oxide.**			
P 103	—, **1,2-epoxy-3-(o-isopropylphenoxy)-,** polyether			
P 104	—, **1,2-epoxy-2-methyl-,** polyether			

No.	$\dfrac{10^4 \, d\bar{v}/dT}{\text{cc g}^{-1} \text{ deg}^{-1}}$	$\dfrac{d_a}{\text{g cm}^{-3}}$	$\dfrac{d_c}{\text{g cm}^{-3}}$	$\dfrac{T_g}{°C}$	$\dfrac{T_d}{°C}$	$\dfrac{T_m}{°C}$	References
P 74						284	410
P 75						42	384
P 76				−60			51
P 77			$(1.0210)^{20}$			50	330
P 78						51	358, (195)
P 79						67	384
P 80						115	157
P 81						168	157
P 82			(1.32)	53		220	420, 347
P 83						70	340
P 84						100	340
P 85			(1.35)	60		245	420, 347
P 86					$205_{0.1}$		455
P 87					$295_{0.1}$		455
P 88						270	210
P 89				70s			199
P 90		(1.32)	1.36				302
P 91							
P 92	1.118^{30}			−64s			461
P 93				−88s			461
P 94						293	521
P 95							
P 96							
P 97						180	259
P 98						126	259
P 99	$(1.354)^{23}$		1.414			200	521
P 100						176	521
P 101							
P 102							
P 103						128	521
P 104						160	143, (259)

No.	Polymer	$\dfrac{\delta}{\mathrm{cal}^{1/2}\,\mathrm{cm}^{-3/2}}$	$\dfrac{\varDelta H_u}{\mathrm{cal\,mol^{-1}}}$	n
P 105	**Propane, 1,2-epoxy-3-(1-naphthyloxy)-**, polyether			
P 106	**–, 1,2-epoxy-3-(2-naphthyloxy)-**, polyether			
P 107	**–, 1,2-epoxy-3-phenoxy-**, polyether			
P 108	**–, 1,2-epoxy-3-(*m*-tolyloxy)-**, polyether			
P 109	**–, 1,2-epoxy-3-(*o*-tolyloxy)-**, polyether			
P 110	**–, 1,2-epoxy-3-(*p*-tolyloxy)-**, polyether			
P 111	**–, 1,2-epoxy-3-(2,4,6-trichlorophenoxy)-**, polyether			
P 112	**Propanedioic acid.** See **Malonic acid.**			
P 113	**Propanoic acid.** See **Propionic acid.**			
P 114	**2-Propanone.** See **Acetone.**			
P 115	**Propenamide.** See **Acrylamide.**			
P 116	**Propene**			
P 117	**–,** (isotactic)		1470	
			1890	
			2000	
			2310	
			2600	
P 118	**–,** (syndiotactic)		450	
P 119	**–, hexafluoro-,** (isotactic)			
P 120	**–, 2-methyl-**		7.7	1.5045_{D}^{27}
		7.8		1.508^{20}
		8.05		
		8.1		
P 121	**–, 2-phenyl-.** See **Styrene, α-methyl-.**			
P 122	**–, 3-phenyl-.** See **Benzene, allyl-.**			
P 123	**–, 3,3,3-trifluoro-**			
P 124	**Propene-2-carboxylic acid.** See **Methacrylic acid.**			
P 125	**Propenenitrile.** See **Acrylonitrile.**			

124

No.	$\dfrac{10^4\,d\bar{v}/dT}{\text{cc g}^{-1}\,\text{deg}^{-1}}$	$\dfrac{d_a}{\text{g cm}^{-3}}$	$\dfrac{d_c}{\text{g cm}^{-3}}$	$\dfrac{T_g}{°C}$	$\dfrac{T_d}{°C}$	$\dfrac{T_m}{°C}$	References
P 105						235	521
P 106						297	521
P 107		1.272^{20}	1.305^{20}	42	35_{138}	215	376, (242), (521)
		(1.21)					376
P 108						169	521
P 109						191	521
P 110						212	521
P 111						130	521
P 112							
P 113							
P 114							
P 115							
P 116	2.2/9.4	0.858^{25}		40.6	0_{1000}		336, 445, 41
				−9.4	-38_{1500}		(114), (76), (286)
P 117	2.2/9.4	0.852^{30}	0.943α	26	27_{1000}	183α	114, 336, 505, 41, 408
	/9.03	0.766^{176}	0.920β	−35	-38_{1500}	130β	286, 287, 47
			0.95γ		$-220_{7.15}$	150γ	412, 472, (193), (40)
							397, (46), (237), (407)
							23, (445), (97), (60)
P 118		0.858^{21}	0.90		0_1	165 I	538, 539, 193
			0.898			112 II	539, 538
P 119				11		160	2, 447
P 120	<1/5.6	0.918^{20}	0.956	−76		1.5	7, 189, 562, 54
	/6.0	0.913^{25}					2, 172, 326, 145, 61,
		0.873^{100}					9, 189
		0.835^{178}					182, 189
P 121							
P 122							
P 123		1.580^{20}		55s			479
P 124							
P 125							

No.	Polymer	$\dfrac{\delta}{\mathrm{cal}^{1/2}\,\mathrm{cm}^{-3/2}}$	$\dfrac{\Delta H_u}{\mathrm{cal\ mol}^{-1}}$	n
P 126	**Propenoic acid.** See **Acrylic acid.**			
P 127	**Propionaldehyde,** polyacetal (isotactic)			
P 128	**Propionic acid,** vinyl ester	8.85		1.4665_D^{20}
P 129	—, 3-**amino**-2,2-**dimethyl**-, polyamide		3050 II	
			1870 I	
P 130	—, 3-**chloro**-, vinyl ester			
P 131	—, 3-(p-**chlorophenyl**)-3-**hydroxy**-2,2-**dimethyl**-. See **Hydracrylic acid,** 3-(p-**chlorophenyl**)-2,2-**dimethyl**-.			
P 132	—, 3-**cyclobutyl**-. See **Cyclobutanepropionic acid.**			
P 133	—, 3,3'-(**ethylenedithio**)**di**-, polyanhydride			
P 134	—, 3,3'-**fluorene**-9-**ylidenedi**-. See **Fluoren**-9,9-**dipropionic acid.**			
P 135	—, 3,3'-**furandiyldi**-. See **Furandipropionic acid.**			
P 136	—, 3-(2-**furyl**)-3-**hydroxy**-2,2-**dimethyl**-. See **Hydracrylic acid,** 3-(2-**furyl**)-2,2-**dimethy**			
P 137	—, 3-**hydroxy**-. See **Hydracrylic acid.**			
P 138	—, 3-**hydroxy**-2,2-**dimethyl**-3-(*m*-**nitrophenyl**)-.			
	See **Hydracrylic acid,** 2,2-**dimethyl**-3-(*m*-**nitrophenyl**)-.			
P 139	—, **pentafluoro**-, vinyl ester			1.364_D^{25}
P 140	—, 3,3'-[*p*-**phenylenebis**(**methylenethio**)]**di**-, polyanhydride			
P 141	—, 3,3'-*p*-**phenylenedi**-, polyamide with 1,10-decanediamine			
P 142	—, —, polyamide with 1,12-dodecanediamine			
P 143	—, —, polyamide with 1,6-hexanediamine			
P 144	—, —, polyamide with 1,18-octadecanediamine			
P 145	—, —, polyamide with 1,8-octanediamine			
P 146	—, —, polyamide with 1,14-tetradecanediamine			
P 147	—, —, polyanhydride			
P 148	—, 3,3'-(*p*-**phenylenedithio**)**di**-, polyanhydride			
P 149	—, 3,3'-(**phenylphosphinothioylidene**)**di**-, polyamide with 1,6-hexanediamine			
P 150	—, 3,3'-(**phenylphosphinylidene**)**di**-, polyamide with 1,6-hexanediamine			
P 151	—, 3,3'-**pyrrolidinediyldi**-. See **Pyrrolidinedipropionic acid.**			

126

No.	$\dfrac{10^4\,d\bar{v}/dT}{\text{cc g}^{-1}\text{ deg}^{-1}}$	$\dfrac{d_a}{\text{g cm}^{-3}}$	$\dfrac{d_c}{\text{g cm}^{-3}}$	$\dfrac{T_g}{\text{°C}}$	$\dfrac{T_d}{\text{°C}}$	$\dfrac{T_m}{\text{°C}}$	References
P 126							
P 127			1.05			185	104, 405
P 128		1.02		10	$12_{1.5}$		467, 326, 97, 16
					$-40_{5.4}$		16
P 129			(1.14)II			273II	566
			(1.19) I			189 I	566
P 130					50_{1000}		450
					-109_{1000}		450
P 131							
P 132							
P 133	$(1.386)^{20}$					75	239, 229
P 134							
P 135							
P 136							
P 137							
P 138							
P 139				42			359
P 140						91	229, 239
P 141			$(1.120)^{25}$			271	529
P 142			$(1.080)^{25}$		85_{110}	263	529
P 143						290	157
P 144			$(1.034)^{25}$		65_{110}	232	529
P 145			$(1.151)^{25}$			289	529
P 146			$(1.067)^{25}$			246	529
P 147						92	229
P 148	1.473^{20}					55	239, 229
P 149				43			153
P 150				29			153
P 151							

No.	Polymer	$\dfrac{\delta}{\text{cal}^{1/2}\,\text{cm}^{-3/2}}$	$\dfrac{\Delta H_u}{\text{cal mol}^{-1}}$	n
P 152	**Propionic acid, 3,3′-thiodi-**, polyamide with 1,6-hexanediamine			
P 153	—, **3,3′-thiodi-**, polyanhydride			
P 154	—, **3,3′-(2,5-thiophenediyl)di-**. See **2,5-Thiophenedipropionic acid.**			
P 155	**Propylene.** See **Propene.**			
P 156	**Propylene oxide,** polyether	8.95		1.4495_D^{20}
				1.4515_D^{20}
P 157	—, polyether(isotactic)	9.9	1800	
P 158	**Propylene sulfide,** poly(thioether)			
P 159	**2-Propyn-1-ol,** methacrylate. See **Methacrylic acid,** 2-propynyl ester.			
P 160	**Pyridine, 2-styryl-,** copolymer *(alt)* with 2-vinylpyridine			
P 161	—, **2-vinyl-**			
P 162	—, —, (isotactic)			
P 163	**Pyromellitic acid,** polyimide with 4,4-dimethyl-1,7-heptanediamine			
P 164	—, polyimide with 3-methyl-1,7-heptanediamine			
P 165	—, polyimide with 1,9-nonanediamine			
P 166	**2,5-Pyrroledipropionic acid, 1-methyl-,** polyanhydride			
P 167	**2,5-Pyrrolidinedipropionic acid,** polyanhydride			
P 168	**2-Pyrrolidinone,** polyamide. See **Butyric acid, 4-amino-,** polyamide.			
P 169	—, **1-vinyl-**			1.53_D^{35}
Q 1	**Quadric acid.** See **Benzoic acid, 4,4′-*sec*-butylidenedi-.**			
Q 2	*p*-**Quinodimethan.** See **Benzene,** 1,4-ethylene polymer *(alt)*.			
Q 3	—, **methyl.** See **Toluene,** 2,5-ethylene polymer *(alt)*.			

No.	$\dfrac{10^4 \, d\bar{v}/dT}{\text{cc g}^{-1} \text{deg}^{-1}}$	$\dfrac{d_a}{\text{g cm}^{-3}}$	$\dfrac{d_c}{\text{g cm}^{-3}}$	$\dfrac{T_g}{°C}$	$\dfrac{T_d}{°C}$	$\dfrac{T_m}{°C}$	References
P 152				27		216	153
P 153		$(1.532)^{20}$				55	239, 229
P 154							
P 155							
P 156				-72	-93_{9300}		367, 316, 407
					-183_{10400}		316, 407, (317)
P 157	/7.0	1.002	1.157	-72	$-68_{0.3}$	75	526, 541, 349, 406, 265, 315
	/7.3		1.154		$-150_{0.59}$		266, 314, (532)
			$(1.212)^{20}$				314, (367), (317), (143)
P 158				-47			461
P 159							
P 160				150		350	489
P 161				104			382, (33)
P 162						215	135
P 163				135		320	149, 365
P 164				135			149
P 165				110			149
P 166						188	229
P 167						103	229
P 168							
P 169		1.25^{25}		145			199, 430

Q 1

Q 2

Q 3

No.	Polymer	$\dfrac{\delta}{\mathrm{cal}^{1/2}\,\mathrm{cm}^{-3/2}}$	$\dfrac{\varDelta H_u}{\mathrm{cal\ mol}^{-1}}$	n
S 1	**Salicylic acid,** methyl ester, methacrylate. See **Methacrylic acid,** ester with methyl salicylate.			
S 2	−, phenyl ester, methacrylate. See **Methacrylic acid,** ester with phenyl salicylate.			
S 3	**Sebacic acid,** polyamide with 1,4-butanediamine			
S 4	−, polyamide with *cis*-1,4-cyclohexanebis(methylamine)			
S 5	−, polyamide with *trans*-1,4-cyclohexanebis(methylene)			
S 6	−, polyamide with 1,1'-decamethylenedipiperazine	16500		
S 7	−, polyamide with 1,10-decanediamine	7800		
		8200		
		8300		
S 8	−, polyamide with 1,12-dodecanediamine			
S 9	−, polyamide with ethylenediamine			
S 10	−, polyamide with fluorene-9,9-bis(propylamine)			
S 11	−, polyamide with 1,7-heptanediamine			
S 12	−, polyamide with 1,6-hexanediamine	7300	$1.475(\alpha)$	
		12100	$1.525(\beta)$	
		13050	$1.565(\gamma)$	
		13500		
S 13	−, polyamide with methylenediamine			
S 14	−, polyamide with 1,9-nonanediamine			
S 15	−, polyamide with 1,18-octadecanediamine			
S 16	−, polyamide with 1,8-octanediamine			
S 17	−, polyamide with 1,5-pentanediamine			
S 18	−, polyamide with 2,2'-(*p*-phenylene)bisethylamine			
S 19	−, polyamide with 3,3'-(phenylphosphinidene)bispropylamine			
S 20	−, polyamide with piperazine	6200		
		6300		

No.	$10^4\,\mathrm{d\bar{v}/dT}$ $\mathrm{cc\ g^{-1}\ deg^{-1}}$	d_a $\mathrm{g\ cm^{-3}}$	d_c $\mathrm{g\ cm^{-3}}$	T_g °C	T_d °C	T_m °C	References
S 1							
S 2							
S 3			$(1.100)^{25}$		-36_{133}	250	162, 122, (55)
					-114_{179}		(162), (157), (352)
S 4						208	410
S 5						300	410
S 6	/6.7	0.971^{150}				129.5	307
S 7	/6.7	0.901^{216}	$(1.063)^{25}$	60	67_{600}	216	428, 307, 529, 122, 69, 130
					-42_{194}		162, (4), (157)
					-95_{1200}		69, (348), (352)
S 8			$(1.050)^{25}$			192	529, (352), (55)
S 9					$65_{5.5}$	254	16, 157, (85)
					-55_8		16, (348)
					$-120_{9.8}$		16
S 10				85s			432
S 11						187	352, (348)
S 12		1.041	1.211α	50	$60_{4.0}$	226 I	428, 563, 424, 122, 16
			1.19β		$-60_{5.5}$	165 II	23, 308, (529), (564)
					$-130_{6.5}$		(69), (348), (55),
							570, (157), (85), (352)
S 13						268	354
S 14						176	352, (85)
S 15			$(1.024)^{25}$			171	529
S 16			$(1.075)^{25}$	60		210	529, 122, (55), (348), (155),
							(352), (157)
S 17						195	352
S 18			$(1.139)^{25}$		105_{110}	300	529, (155)
S 19		1.102					153
S 20	/5.9	0.926^{189}		82		183	37, 307, 36
							4, (85)

9*

No.	Polymer	$\dfrac{\delta}{\mathrm{cal}^{1/2}\,\mathrm{cm}^{-3/2}}$	$\dfrac{\Delta H_u}{\mathrm{cal\ mol}^{-1}}$	n
S 21	**Sebacic acid,** polyamide with spiro[3·3]heptane-2,6-diamine			
S 22	—, polyamide with 1,14-tetradecanediamine			
S 23	—, polyamide with 1,11-undecanediamine			
S 24	—, polyamide with p-xylene-α,α'-diamine			
S 25	—, polyanhydride			
S 26	—, polyester with 1,4-butanediol		480	
S 27	—, polyester with *cis*-1,4-cyclohexanedimethanol			
S 28	—, polyester with *trans*-1,4-cyclohexanedimethanol			
S 29	—, polyester with 1,10-decanediol		12000	
			12300	
S 30	—, polyester with diethylene glycol			
S 31	—, polyester with 2,2-dimethyl-1,3-propanediol			
S 32	—, polyester with ethylene glycol		3300	
			6960	
S 33	—, polyester with 1,6-hexanediol			
S 34	—, polyester with isopropylidenediphenol			
S 35	—, polyester with 2-methylene-1,3-propanediol			
S 36	—, polyester with 2-methyl-1,3-propanediol			
S 37	—, polyester with 1,8-octanediol			
S 38	—, polyester with 1,5-pentanediol			
S 39	—, polyester with 1,2-propanediol			
S 40	—, polyester with 1,3-propanediol			
S 41	—, polyester with 4,4'-(thiazolo[5,4-*d*]thiazole-2,5-diyl)bis(2-methoxyphenol)			
S 42	—, polyester with triethylene glycol			
S 43	—, polyester with p-xylene-α,α'-diol			
S 44	**Selenium,** polyselenide			
S 45	**Silane, allyl-,** (isotactic)			
S 46	—, **allyltrimethyl-,** (isotactic)			
S 47	**Silanol,** p-**phenylenebis[dimethyl-,** polysiloxane		4350	

No.	$10^4\,dv/dT$ cc g⁻¹ deg⁻¹	d_a g cm⁻³	d_c g cm⁻³	T_g °C	T_d °C	T_m °C	References
S 21						240	509
S 22			$(1.042)^{25}$			175	529
S 23						169	352
S 24		(1.135)	1.169		115_{110}	300	116, 529, 55, (157), (410)
S 25						83	3
S 26				−57	-35_{74}	62.0	437, 326, 162
					-120_{133}		162, (343)
S 27						50	384
S 28						78	384
S 29	/7.50	0.9378^{109}				84.5	130, 243, 428
							23, 307, (55), (157)
S 30				−74		33	51, 343
S 31						26	358
S 32	1.96/3.58	1.113^{25}	1.211	−30		79	155, 102, 303, 180
		$(1.0354)^{20}$	1.187				330, 411, 195
			1.120				(55), (343)
S 33			$(1.0200)^{20}$			74	330, 55, (157)
S 34				21.5s			371
S 35						42	358
S 36						26	358
S 37						75	57
S 38						51	343
S 39						20	358
S 40			1.090			58	195, 358, (343)
S 41						220	361
S 42						29	343
S 43						93	384
S 44				30			421
S 45						128	381
S 46		(0.863)	(0.876)			360	181, 381
S 47						148	178

No.	Polymer	δ $\mathrm{cal}^{1/2}\,\mathrm{cm}^{-3/2}$	ΔH_u $\mathrm{cal\,mol}^{-1}$	n
S 48	**Silicone, dimethyl-,** polysiloxane	7.5		
		7.3		
S 49	−, **diphenyl-,** polysiloxane			
S 50	−, **methylphenyl-,** polysiloxane			
S 51	**1,1′-Spirobi[indan]-6,6′-diol, 3,3,3′,3′-tetramethyl-,** 2-hydroxytrimethylene polyether *(alt)*			
S 52	**Spiro[3·3]heptane-2,6-dicarboxylic acid,** polyamide with 1,4-butanediamine			
S 53	−, polyamide with 1,10-decanediamine			
S 54	−, polyamide with 1,6-hexanediamine			
S 55	−, polyamide with 2-methylpiperazine			
S 56	−, polyamide with 1,8-octanediamine			
S 57	−, polyamide with piperazine			
S 68	−, polyamide with spiro[3.3]heptane-2,6-diamine			
S 59	−, polyamide with m-xylene-α,α'-diamine			
S 60	**Stearic acid,** vinyl ester			
S 61	**2-Stilbazole.** See **Pyridine, 2-styryl-.**			
S 62	**4,4′-Stilbenedicarboxylic acid,** polyester with ethylene glycol			
S 63	**Styrene**	8.56		1.5916_D^{20}
		9.1		1.5910_D^{20}
		9.12		
S 64	−, (isotactic)		2000	
			2100	
			2400	
S 65	−, p-**bromo**			
S 66	−, **2-bromo-4-(trifluoromethyl)-**			1.5_D^{25}
S 67	−, p-**butyl-**			
S 68	−, p-*sec*-**butyl-**			
S 69	−, p-*tert*-**butyl-**			

134

No.	$\dfrac{10^4\ dv/dT}{cc\ g^{-1}\ deg^{-1}}$	d_a g cm^{-3}	d_c g cm^{-3}	T_g °C	T_d °C	T_m °C	References
S 48	2.7/12	0.974^{25}		−123		−29	264, 2, 92, 151, (184)
	9	0.976^{25}					2, 518, (61), (1)
S 49						373	150
S 50				−86		−29	184, 151
S 51					$140_{0.1}$		455
S 52					>345		509
S 53				>80	230		509
S 54				110	270		509
S 55					>300		509
S 56					245		509
S 57					>300		509
S 58					>365		509
S 59				160	265		509
S 60						51.7	353, (42)
S 61							
S 62						420	180
S 63	2.5/5.5	1.057^{0}		100	$116_{0.9}$		2, 311, 163, 32, 16
	2.18/6.47	1.047^{25}			70_{8}		9, 480, 77, (61)
	1.9/6.0	1.051^{25}			-140_{11}		7, 460, 481,
	2.2/5.8	0.9615^{217}			$-235_{5.6}$		472, (19)
	2.4/4.32						513
S 64	2.50/5.6	1.053^{25}	1.113^{30}	100	-228_{7100}	240	480, 287, 24, 452, 138
	$\cdot 1.056^{25}$	1.115^{25}					415, 481
	1.053^{30}	1.111					481, 66, 446
		1.126					138
S 65				144	132_{1}		77, 197
S 66		1.71^{33}					304
S 67				6	48_{1}		39, 196, (19)
S 68				86s			19
S 69				132s			96, (19)

No.	Polymer	$\dfrac{\delta}{\mathrm{cal}^{1/2}\,\mathrm{cm}^{-3/2}}$	$\dfrac{\Delta H_u}{\mathrm{cal\ mol}^{-1}}$	n
S 70	**Styrene, 5-*tert*-butyl-2-methyl**			
S 71	**−, *m*-chloro-**			
S 72	**−, *o*-chloro**			1.6098_D^{20}
S 73	**−, *p*-chloro-**			
S 74	**−, 4-chloro-3-fluoro-**			
S 75	**−, 4-chloro-2-methyl-**			
S 76	**−, 4-chloro-3-methyl-**			
S 77	**−, *m*-cyano-**			
S 78	**−, *p*-cyano-**			
S 79	**−, *p*-decyl-**			
S 80	**−, 2,4-dichloro-**			
S 81	**−, 2,5-dichloro-**			
S 82	**−, 2,6-dichloro-**			1.6248_D^{20}
S 83	**−, 3,4-dichloro-**			
S 84	**−, 2,5-difluoro-**			
S 85	**−, 2,4-diisopropyl-**			
S 86	**−, *α-p*-dimethyl-**			
S 87	**−, 2,4-dimethyl-**			
S 88	**−, 2,4-dimethyl-**, (isotactic)			
S 89	**−, 2,5-dimethyl-**			
S 90	**−, −**, (isotactic)			
S 91	**−, 3,4-dimethyl-**			
S 92	**−, 3,4-dimethyl-**, (isotactic)			
S 93	**−, 3,5-dimethyl-**, (isotactic)			
S 94	**−, *p*-dodecyl-**			
S 95	**−, *m*-ethyl-**			
S 96	**−, *o*-ethyl-**			
S 97	**−, *p*-ethyl-**			
S 98	**−, *m*-fluoro-**, (stereoregular)			
S 99	**−, *o*-fluoro-**			

No.	$\dfrac{10^4 \, dv/dT}{cc \ g^{-1} \ deg^{-1}}$	$\dfrac{d_a}{g \ cm^{-3}}$	$\dfrac{d_c}{g \ cm^{-3}}$	$\dfrac{T_g}{°C}$	$\dfrac{T_d}{°C}$	$\dfrac{T_m}{°C}$	References
S 70				87			77
S 71				90			77
S 72				119			311, 77
S 73				116	125_1		77, 197, (96)
					-178_{8330}		556
S 74				122			77
S 75				145			77
S 76				114			77
S 77				100s			199
S 78	2.2/4.6			120			460, (199)
S 79				−65			39
S 80				135			77
S 81				120			14, (96), (199), (77)
S 82				167s			311, 96
S 83				128			77, (96)
S 84				101s			96
S 85				168s			96
S 86					-110_{8980}		556
S 87				121			24, (96), (77), (19)
S 88				100		350	295, 195, (446)
S 89				143			77, (96)
S 90						340	195, (295), (446)
S 91				111			77, (96)
S 92				100		240	295, 446
S 93				104		290	295, 446
S 94				−52			39
S 95				30s			19
S 96				103s			19
S 97				78			196, (39), (19)
S 98						250	94
S 99				105s			199

No.	Polymer	$\dfrac{\delta}{\text{cal}^{1/2}\ \text{cm}^{-3/2}}$	$\dfrac{\Delta H_u}{\text{cal mol}^{-1}}$	n
S 100	**Styrene, *o*-fluoro-**, (isotactic)			
S 101	−, ***p*-fluoro-**			
S 102	−, −, (isotactic)			
S 103	−, **4-fluoro-2-methyl-**, (isotactic)			
S 104	−, **4-fluoro-2-(trifluoromethyl)-**			1.46_D^{25}
S 105	−, **4-fluoro-3-(trifluoromethyl)-**			1.46_D^{25}
S 106	−, ***p*-hexadecyl-**			
S 107	−, ***p*-hexyl-**			
S 108	−, ***p*-iodo-**			
S 109	−, ***p*-isopropyl-**			1.554_D^{20}
S 110	−, ***o*-methoxy-**			1.5932_D^{20}
S 111	−, −, (stereoregular)			
S 112	−, ***p*-methoxy-**			1.5967_D^{20}
S 113	−, −, (stereoregular)			
S 114	−, ***p*-methoxycarbonyl-**			
S 115	−, ***p*-methoxy-β-methyl-**			
S 116	−, **4-methoxy-2-methyl-**			1.5868_D^{20}
S 117	−, **α-methyl-**			
S 118	−, ***m*-methyl-**			
S 119	−, −, (isotactic)			
S 120	−, ***o*-methyl-**			1.5874_D^{20}
S 121	−, −, (isotactic)			
S 122	−, ***p*.methyl-**			
S 123	−, ***p*-nonyl-**			
S 124	−, ***p*-octadecyl-**			
S 125	−, ***p*-octyl-**			
S 126	−, ***p*-phenoxy-**. See **Ether, phenyl *p*-vinylphenyl.**			
S 127	−, ***p*-phenyl-**. See **Biphenyl, 4-vinyl-.**			

138

No.	$\dfrac{10^4 \ dv/dT}{cc \ g^{-1} \ deg^{-1}}$	$\dfrac{d_a}{g \ cm^{-3}}$	$\dfrac{d_c}{g \ cm^{-3}}$	$\dfrac{T_g}{°C}$	$\dfrac{T_d}{°C}$	$\dfrac{T_m}{°C}$	References
S 100			1.29			270	139, 76, 415, 446
			1.526				195
S 101				106	109_1		77, 197
S 102				80		265	295, 446, (94)
S 103					>360		446, (76)
S 104		1.43^{33}					304
S 105		1.43^{33}					304
S 106				4.5			39
S 107				−27			39
S 108					156_1		197
S 109				87s			311, 19
S 110				80s			311
S 111				110s			448
S 112				60s	-153_{7060}		311, 556
S 113			(1.12)			238	133, (448)
S 114	2.2/5.4			131			460
S 115					-153_{7220}		556
S 116				90s			311
S 117		1.066^{25}		170	-133_{6700}		452, 480, (96), (1)
		1.063					332, (556)
S 118				97			77, (96), (19)
S 119			1.010	70		215	195, 295, 268, 446
			1.02				268
S 120		1.027^{24}		135.5	-230_{7000}		311, 452, 77, (96), (19)
S 121			1.07	96		>360	140, 295, 446
S 122		1.04		106	-181_{8700}		444, 24, 556, (96), (19), (77)
S 123				−53			39
S 124				32s			39
S 125				−45			39
S 126							
S 127							

No.	Polymer	$\dfrac{\delta}{\mathrm{cal}^{1/2}\,\mathrm{cm}^{-3/2}}$	$\dfrac{\Delta H_u}{\mathrm{cal\ mol}^{-1}}$	n
S 128	**Styrene, *p*-propyl-**			
S 129	—, *p*-**tetradecyl-**			
S 130	—, **trichloro-**, (mixed isomers)			
S 131	—, α,β,β-**trifluoro-**			
S 132	—, *m*-**(trifluoromethyl)-**			
S 133	—, 2,4,5-trimethyl-			
S 134	—, 2,4,6-trimethyl-			
S 135	**Styrene oxide.** See **Benzene, (epoxyethyl)-.**			
S 136	**Styrene sulfide.** See **Benzene, (epithioethyl)-.**			
S 137	**Suberic acid,** polyamide with 1,4-butanediamine			
S 138	—, polyamide with *cis*-1,4-cyclohexanebis(methylamine)			
S 139	—, polyamide with *trans*-1,4-cyclohexanebis(methylamine)			
S 140	—, polyamide with 1,10-decanediamine			
S 141	—, polyamide with 1,7-heptanediamine			
S 142	—, polyamide with 1,6-hexanediamine			
S 143	—, polyamide with 1,8-octanediamine			
S 144	—, polyamide with 1,5-pentanediamine			
S 145	—, polyamide with spiro[3.3]heptane-2,6-diamine			
S 146	—, polyamide with *p*-xylene-α,α'-diamine			
S 147	—, polyester with *cis*-1,4-cyclohexanedimethanol			
S 148	—, polyester with *trans*-1,4-cyclohexanedimethanol			
S 149	—, polyester with diethylene glycol			
S 150	—, polyester with 2,2-dimethyl-1,3-propanediol			
S 151	—, polyester with ethylene glycol			
S 152	—, polyester with 1,3-propanediol			
S 153	—, polyester with *p*-xylene-α,α'-diol			
S 154	**Succinic acid,** polyester with 2,2-bis(chloromethyl)-1,3-propanediol			
S 155	—, polyester with 1,4-butanediol			
S 156	—, polyester with *cis*-1,4-cyclohexanedimethanol			

140

No.	$\dfrac{10^4 \, dv/dT}{\text{cc g}^{-1}\text{ deg}^{-1}}$	$\dfrac{d_a}{\text{g cm}^{-3}}$	$\dfrac{d_c}{\text{g cm}^{-3}}$	$\dfrac{T_g}{^\circ\text{C}}$	$\dfrac{T_d}{^\circ\text{C}}$	$\dfrac{T_m}{^\circ\text{C}}$	References
S 128				25		39	
S 129				−36		39	
S 130				145s		96	
S 131				240s		96	
S 132	1.32^{33}					304	
S 133				136s		96	
S 134				162s		96	
S 135							
S 136							
S 137					250		55, 157, 352
S 138					215		410
S 139					311		410
S 140					217s		348
S 141					230s		348
S 142					$64_{3.1}$	235	16, 157, 57, (348)
					$-50_{5.4}$		16, (352), (55)
					$-120_{6.8}$		16
S 143			$(1.090)^{25}$		-39_{198}	225s	162, 348, (55), (352)
S 144					202		352
S 145					295		509
S 146					305		410
S 147					50		384
S 148					96		384
S 149				−61			51
S 150					17		358
S 151					45		57
S 152					52		358, (195)
S 153					82		384
S 154					74		462
S 155					118		358, (343)
S 156					62		384

No.	Polymer	$\dfrac{\delta}{\text{cal}^{1/2}\,\text{cm}^{-3/2}}$	$\dfrac{\Delta H_u}{\text{cal mol}^{-1}}$	n
S 157	**Succinic acid,** polyester with *trans*-1,4-cyclohexanedimethanol			
S 158	—, polyester with 1,10-decanediol			
S 159	—, polyester with diethylene glycol			
S 160	—, polyester with 2,2-diethyl-1,3-propanediol			
S 161	—, polyester with 2,2-dimethyl-1,3-propanediol			
S 162	—, polyester with ethylene glycol			
S 163	—, polyester with 1,6-hexanediol			
S 164	—, polyester with hydroquinone			
S 165	—, polyester with 1,3-propanediol			
S 166	—, polyester with 4,4'-((thiozolo[5,4-*d*]thiazole-2,5-diyl)bis(2-methoxyphenol)			
S 167	—, polyester with *p*-xylene-α,α'-diol			
S 168	—, *d*-**2,3-dimethoxy-**, polyester with 1,4-butanediol			
S 169	—, **2-methyl-**, polyester with 1,10-decanediol			
S 170	**Succinimide,** *N*-**vinyl-**			
S 171	**Sulfide, butyl vinyl**			
S 172	—, **decamethylene,** poly(thioether)			
S 173	—, —, poly(thioether) *(alt)* with hexamethylene sulfide			
S 174	—, —, poly(thioether) *(alt)* with tetrahydrothiophene			
S 175	—, —, poly(thieether) *(alt)* with undecamethylene sulfide			
S 176	—, **ethyl vinyl**			
S 177	—, **hexamethylene,** poly(thioether)			
S 178	—, —, poly(thioether) *(alt)* with 4,4'-biphenylylene sulfide			
S 179	—, —, poly(thioether) *(alt)* with 4-hydroxyheptamethylene sulfide			
S 180	—, —, poly(thioether) *(alt)* with methylenebis(*p*-phenyleneëthylene) sulfide			
S 181	—, —, poly(thioether) *(alt)* with methylenedi-*p*-phenylene sulfide			
S 182	—, —, poly(thioether) *(alt)* with 1,5-naphthylene sulfide			
S 183	—, —, poly(thioether) *(alt)* with 2,6-naphthylene sulfide			
S 184	—, —, poly(thioether) *(alt)* with 2,7-naphthylene sulfide			
S 185	—, —, poly(thioether) *(alt)* with 4,4'-oxydiphenylene sulfide			
S 186	—, —, poly(thioether) *(alt)* with pentamethylene sulfide			
S 187	—, —, poly(thioether) *(alt)* with *p*-phenylene sulfide			

No.	$\dfrac{10^4 \, dv/dT}{cc \, g^{-1} \, deg^{-1}}$	$\dfrac{d_a}{g \, cm^{-3}}$	$\dfrac{d_c}{g \, cm^{-3}}$	$\dfrac{T_g}{°C}$	$\dfrac{T_d}{°C}$	$\dfrac{T_m}{°C}$	References
S 157						147	384
S 158						68	55
S 159				−29			51
S 160						46	358
S 161				−18		86	227, 358
S 162	3.16/4.04		1.175	−1		108	102, 195, 55
S 163						57	55
S 164						300s	155
S 165						52	358, (195)
S 166						340	361
S 167						115	384
S 168						92	358
S 169						2	358
S 170				210s			199
S 171	0.984[27]			−20			403
S 172						91.5	120
S 173						78	55
S 174						86	120
S 175						78	241
S 176				−7			403
S 177						79.5	120, (55)
S 178						235	241
S 179						69	241
S 180						88	241
S 181						121	241
S 182						151	241
S 183						190	241
S 184						90	241
S 185						119	241
S 186						65	241
S 187						159	241

No.	Polymer	$\dfrac{\delta}{\mathrm{cal}^{1/2}\,\mathrm{cm}^{-3/2}}$	$\dfrac{\Delta H_u}{\mathrm{cal\,mol}^{-1}}$	n
S 188	**Sulfide, hexamethylene,** poly(thioether) *(alt)* with tetrahydrothiophene			
S 189	−, −, poly(thioether) *(alt)* with undecamethylene sulfide			
S 190	−, **methyl vinyl**			
S 191	−, **methylenebis(*p*-phenyleneëthylene),** poly(thioether)			
S 192	−, **pentamethylene,** poly(thioether) *(alt)* with tetrahydrothiophene			
S 193	−, *p*-**phenylene,** poly(thioether)			
S 194	−, **phenyl vinyl**			1.6568_D^{20}
S 195	**Sulfone, hexamethylene,** polysulfone			
S 196	−, −, polysulfone *(alt)* with pentamethylene sulfone			
S 197	−, −, polysulfone *(alt)* with tetramethylene sulfone			
S 198	−, **pentamethylene,** polysulfone			
S 199	−, −, polysulfone *(alt)* with tetramethylene sulfone			
S 200	−, **phenylethylene,** polysulfone			
T 1	D-**Tartaric acid,** polyester with 1,10-decanediol			
T 2	*meso*-**Tartaric acid,** polyester with 1,10-decanediol			
T 3	**Terephthalamic acid,** N,N'-**ethylenedi-,** polyesteramide with 1,10-decanediol			
T 4	−, −, polyesteramide with 1,12-dodecanediol			
T 5	−, −, polyesteramide with 1,7-heptanediol			
T 6	−, −, polyesteramide with 1,6-hexanediol			
T 7	−, −, polyesteramide with 1,9-nonanediol			
T 8	−, −, polyesteramide with 1,8-octanediol			
T 9	−, −, polyesteramide with 1,5-pentanediol			
T 10	−, −, polyesteramide with 1,14-tetradecanediol			
T 11	−, −, polyesteramide with 1,13-tridecanediol			
T 12	−, −, polyesteramide with 1,11-undecanediol			
T 13	−, N, N'-**hexamethylenedi-,** polyesteramide with 1,4-butanediol			
T 14	−, −, polyesteramide with 1,10-decanediol			
T 15	−, −, polyesteramide with 1,12-dodecanediol			
T 16	−, −, polyesteramide with 1,7-heptanediol			

No.	$\dfrac{10^4 \, dv/dT}{cc \; g^{-1} \, deg^{-1}}$	$\dfrac{d_a}{g \; cm^{-3}}$	$\dfrac{d_c}{g \; cm^{-3}}$	$\dfrac{T_g}{°C}$	$\dfrac{T_d}{°C}$	$\dfrac{T_m}{°C}$	References
S 188						75	120, (55)
S 189						78	241
S 190	1.184^{27}			−1			403
S 191						108	241
S 192						67	120
S 193				204s		285	70
S 194							311
S 195			1.385			220	475, 377
S 196			1.418				475
S 197			1.46				475
S 198			1.475				475
S 199			1.532				475
S 200						215	181
T 1						66	358
T 2						33	358
T 3						277	250
T 4						264	250
T 5						275	250
T 6						274	250
T 7						267	250
T 8						282	250
T 9						312	250
T 10						249	250
T 11						257	250
T 12						251	250
T 13						300	250
T 14						235	250
T 15						231	250
T 16						227	250

No.	Polymer	$\dfrac{\delta}{\mathrm{cal}^{1/2}\,\mathrm{cm}^{-3/2}}$	$\dfrac{\varDelta H_u}{\mathrm{cal\ mol^{-1}}}$	n
T 17	**Terephthalamic acid,** N,N'-**hexamethylenedi-,** polyesteramide with 1,6-hexanediol			
T 18	−, −, polyesteramide with 1,9-nonanediol			
T 19	−, −, polyesteramide with 1,8-octanediol			
T 20	−, −, polyesteramide with 1,5-pentanediol			
T 21	−, −, polyesteramide with 1,14-tetradecanediol			
T 22	−, −, polyesteramide with 1,13-tridecanediol			
T 23	−, −, polyesteramide with 1,11-undecanediol			
T 24	−, N,N'-**methylenedi-,** polyanhydrideamide			
T 25	−, N,N'-**pentamethylenedi-,** polyesteramide with 1,10-decanediol			
T 26	−, −, polyesteramide with 1,12-dodecanediol			
T 27	−, −, polyesteramide with 1,7-heptanediol			
T 28	−, −, polyesteramide with 1,6-hexanediol			
T 29	−, −, polyesteramide with 1,9-nonanediol			
T 30	−, −, polyesteramide with 1,8-octanediol			
T 31	−, −, polyesteramide with 1,5-pentanediol			
T 32	−, −, polyesteramide with 1,14-tetradecanediol			
T 33	−, −, polyesteramide with 1,13-tridecanediol			
T 34	−, −, polyesteramide with 1,11-undecanediol			
T 35	−, N,N'-**tetramethylenedi-,** polyesteramide with 1,10-decanediol			
T 36	−, −, polyesteramide with 1,12-dodecanediol			
T 37	−, −, polyesteramide with 1,7-heptanediol			
T 38	−, −, polyesteramide with 1,6-hexanediol			
T 39	−, −, polyesteramide with 1,9-nonanediol			
T 40	−, −, polyesteramide with 1,8-octanediol			
T 41	−, −, polyesteramide with 1,5-pentanediol			
T 42	−, −, polyesteramide, with 1,14-tetradecanediol			
T 43	−, −, polyesteramide with 1,13-tridecanediol			
T 44	−, −, polyesteramide with 1,11-undecanediol			
T 45	−, N,N'-**trimethylenedi-,** polyesteramide with 1,4-butanediol			
T 46	−, −, polyesteramide with 1,10-decanediol			
T 47	−, −, polyesteramide with 1,12-dodecanediol			

146

No.	$\dfrac{10^4\,dv/dT}{cc\ g^{-1}\ deg^{-1}}$	$\dfrac{d_a}{g\ cm^{-3}}$	$\dfrac{d_c}{g\ cm^{-3}}$	$\dfrac{T_g}{{}^\circ C}$	$\dfrac{T_d}{{}^\circ C}$	$\dfrac{T_m}{{}^\circ C}$	References
T 17						265	250
T 18						219	250
T 19						238	250
T 20						255	250
T 21						222	250
T 22						222	250
T 23						217	250
T 24						330	229
T 25						205	250
T 26						206	250
T 27						199	250
T 28						238	250
T 29						194	250
T 30						207	250
T 31						198	250
T 32						200	250
T 33						194	250
T 34						199	250
T 35						261	250
T 36						258	250
T 37						269	250
T 38						290	250
T 39						259	250
T 40						276	250
T 41						284	250
T 42						248	250
T 43						248	250
T 44						248	250
T 45						193	250
T 46						222	250
T 47						216	250

No.	Polymer	$\dfrac{\delta}{\mathrm{cal}^{1/2}\,\mathrm{cm}^{-3/2}}$	$\dfrac{\Delta H_u}{\mathrm{cal}\,\mathrm{mol}^{-1}}$	n
T 48	**Terephthalamic acid, N,N'-trimethylenedi-,** polyesteramide with ethylene glycol			
T 49	$-$, $-$, polyesteramide with 1,7-heptanediol			
T 50	$-$, $-$, polyesteramide with 1,6-hexanediol			
T 51	$-$, $-$, polyesteramide with 1,9-nonanediol			
T 52	$-$, $-$, polyesteramide with 1,8-octanediol			
T 53	$-$, $-$, polyesteramide with 1,5-pentanediol			
T 54	$-$, $-$, polyesteramide with 1,3-propanediol			
T 55	$-$, $-$, polyesteramide with 1,14-tetradecanediol			
T 56	$-$, $-$, polyesteramide with 1,13-tridecanediol			
T 57	$-$, $-$, polyesteramide with 1,11-undecanediol			
T 58	**Terephthalic acid,** polyamide with 1,4-butanediamine			
T 59	$-$, polyamide with N,N'-dimethyl-1,4-butanediamine			
T 60	$-$, polyamide with N,N'-dimethylethylenediamine			
T 61	$-$, polyamide with 1,12-dodecanediamine			
T 62	$-$, polyamide with ethylenediamine			
T 63	$-$, polyamide with 1,7-heptanediamine			
T 64	$-$, polyamide with 1,6-hexanediamine			
T 65	$-$, polyamide with 1,18-octadecanediamine			
T 66	$-$, polyamide with 3,3'-oxybispropylamine			
T 67	$-$, polyamide with 1,5-pentanediamine			
T 68	$-$, polyamide with N,N'-m-phenylenebis(m-aminobenzamide)			
T 69	$-$, polyamide with N,N'-m-phenylenebis(p-aminobenzamide)			
T 70	$-$, polyamide with N,N'-p-phenylenebis(m-aminobenzamide)			
T 71	$-$, polyamide with N,N'-p-phenylenebis(p-aminobenzamide)			
T 72	$-$, polyamide with m-phenylenediamine			
T 73	$-$, polyamide with p-phenylenediamine			
T 74	$-$, polyamide with piperazine			
T 75	$-$, polyamide with 1,3-propanediamine			
T 76	$-$, polyamide with 1,14-tetradecanediamine			
T 77	$-$, polyamide with 1,11-undecanediamine			
T 78	$-$, polyanhydride			

No.	$10^4\ dv/dT$ cc g^{-1} deg^{-1}	d_a g cm^{-3}	d_c g cm^{-3}	T_g °C	T_d °C	T_m °C	References
T 48						205	250
T 49						222	250
T 50						256	250
T 51						211	250
T 52						231	250
T 53						235	250
T 54						235	250
T 55						208	250
T 56						206	250
T 57						198	250
T 58						436	207
T 59						272	207
T 60						379	207
T 61			$(1.152)^{25}$			301	529, 75
T 62						455	207
T 63					123	328	252
T 64						371	207, (155), (57), (157)
T 65			$(1.067)^{25}$			255	529
T 66					72	281	252
T 67						353	207
T 68	(1.36)					450	507
T 69						490	507
T 70				295		467	507
T 71						555	507
T 72						500	249
T 73						500	249
T 74						>420	420, (85)
T 75						399	207
T 76			$(1.115)^{25}$			265	529
T 77						292	75
T 78						400	229, (155)

No.	Polymer	$\dfrac{\delta}{\text{cal}^{1/2}\,\text{cm}^{-3/2}}$	$\dfrac{\varDelta H_u}{\text{cal mol}^{-1}}$	n
T 79	**Terephthalic acid,** polyester with 4,4'-benzylidenediphenol			
T 80	—, polyester with 3,3-bis(p-hydroxyphenyl)-2-methylphthalimidine			
T 81	—, polyester with 3,3-bis(p-hydroxyphenyl)phthalimidine			
T 82	—, polyester with 1,3-butanediol			
T 83	—, polyester with 1,4-butanediol		7600	
T 84	—, polyester with cis-1,4-cyclohexanedimethanol			
T 85	—, polyester with $trans$-1,4-cyclohexanedimethanol			
T 86	—, polyester with cis-1,2-cyclopropanedimethanol			
T 87	—, polyester with $trans$-1,2-cyclopropanedimethanol			
T 88	—, polyester with 1,10-decanediol		11000	
T 89	—, polyester with 2,2'-(2,5-di-$tert$-butyl-p-phenylenedioxy)diethanol			
T 90	—, polyester with 2,2-dimethyl-1,3-propanediol			
T 91	—, polyester with 2,2'-dipropyl-p,p'-biphenol			
T 92	—, polyester with ethylene glycol	10.7	2200	
		9.7	5400	
			5820	
T 93	—, polyester with 4,4'-(1-ethylpropylidene)diphenol			
T 94	—, polyester with 1,7-heptanediol			
T 95	—, polyester with 1,6-hexanediol		8300	
			8400	
			8500	
T 96	—, polyester with 4,4'-isopropylidenediphenol			
T 97	—, polyester with 1,9-nonanediol			
T 98	—, polyester with 1,18-octadecanediol			
T 99	—, polyester with 1,8-octanediol			

No.	$\dfrac{10^4\ dv/dT}{cc\ g^{-1}\ deg^{-1}}$	$\dfrac{d_a}{g\ cm^{-3}}$	$\dfrac{d_c}{g\ cm^{-3}}$	$\dfrac{T_g}{°C}$	$\dfrac{T_d}{°C}$	$\dfrac{T_m}{°C}$	References
T 79				220s			154
T 80				282			362
T 81				327			362
T 82				80			157
T 83		1.08^{230}		22	44_1	232	251, 537, 269, 79
					-100_1		269, (155), (55)
					-130_1		
T 84			1.303			256	185, 384
T 85		1.19	1.265	92		318	185, 174, 384
T 86						130	340
T 87						130	340
T 88	/5.3	1.050^{120}	$(1.097)^{120}$	-5	25_{100}	138	130, 307, 306, 537, 79
		1.032^{150}			-125_{220}		23, (55)
		1.046^{150}					307
T 89						227	236
T 90						140	131, 157
T 91						350	371
T 92	2.23/2.92	1.339^0	1.46^{25}	69	$81_{1.2}$	270	2, 155, 102, 32, 38, 22
	1.40/7.37	1.33^{25}	1.455		$-70_{1.2}$		263, 23, 167, 270, 7
	2.4/6.0	1.331^{30}	1.47		-105_1		195, 553, 128, 269, (537)
		1.334^{30}			-165_1		263, 269
							(55), (79)
T 93				180s			154
T 94				3		98	537
T 95		1.166^{180}	1.131	-9	45_{160}	160.5	130, 307, 195, 537, 79
					-110_{380}		23, (155), (55)
							307
T 96				214		350	115, 371
T 97				-3	35_{370}	90	537, 79
T 98						116	55
T 99					45_{350}	132	79, 55, 537
					-100_{330}		79

No.	Polymer	$\dfrac{\delta}{cal^{1/2}\ cm^{-3/2}}$	$\dfrac{\Delta H_u}{cal\ mol^{-1}}$	n
T 100	**Terephthalic acid,** polyester with 1,5-pentanediol			
T 101	—, polyester with 1,2-propanediol			
T 102	—, polyester with 1,3-propanediol			
T 103	—, polyester with 4,4′-(2,2,2-trichloroethylidene)diphenol			
T 104	—, polyester with p-xylene-α,α'-diol			
T 105	—, **2,5-dichloro-,** polyester with ethylene glycol			
T 106	—, **2,5-dimethyl-,** polyester with ethylene glycol			
T 107	—, **methyl-,** polyester with ethylene glycol			
T 108	**Tetradecanedioic acid,** polyamide with 1,5-pentanediamine			
T 109	—, polyamide with 2,2′-p-phenylenebis(ethylamine)			
T 110	—, polyamide with p-xylene-α,α'-diamine			
T 111	**Tetradecanoic acid. See Myristic acid.**			
T 112	**1-Tetradecene,** (isotactic)			
T 113	**2,4,8,10-Tetroxaspiro[5.5]undecane-3,9-dicarboxylic acid,** polyester with ethylene glycol			
T 114	**6-Thiabicyclo[3.1.1]heptane,** poly(thioether)			
T 115	—, **6,6-dioxide-,** polysulfone			
T 116	**7-Thiabicyclo[4.1.0]heptane,** poly(thioether)			
T 117	—, **7,7-dioxide-,** polysulfone			
T 118	**2-Thiaspiro[3.5]nonane, 2,2-dioxide,** polysulfone			
T 119	**Thiazole, 5,5′-p-phenylenedi-,** 2,2′-m-phenylene polymer *(alt)*			
T 120	—, —, 2,2′-p-phenylene polymer *(alt)*			
T 121	—, —, 2,2′-tetramethylene polymer *(alt)*			
T 122	**Thietane,** poly(thioether)			
T 123	—, poly(thioether) *(alt)* with tetrahydrothiophene			
T 124	—, **3,3-diethyl-, 1,1-dioxide,** polysulfone			
T 125	—, **3,3-dimethyl-,** poly(thioether)			
T 126	—, —, **1,1-dioxide,** polysulfone			

152

No.	$\dfrac{10^4\ dv/dT}{\text{cc g}^{-1}\text{ deg}^{-1}}$	$\dfrac{d_a}{\text{g cm}^{-3}}$	$\dfrac{d_c}{\text{g cm}^{-3}}$	$\dfrac{T_g}{°C}$	$\dfrac{T_d}{°C}$	$\dfrac{T_m}{°C}$	References
T 100				10	45_{210}	134	537, 79, (157)
					-95_{400}		79
T 101				68		122	32, 180
T 102				35	95_{270}	233	537, 79, (131), (155), (157)
					-30_{530}		79
T 103				280s			154
T 104						272	384
T 105						165	180
T 106						180	180
T 107	1.170		(1.199)			70	270, 180
T 108						178	352
T 109			$(1.091)^{25}$		93_{110}	267	529
T 110			$(1.110)^{25}$			257	529
T 111							
T 112					10_{60}	57	97, 372
					-130_{450}		97
T 113						160	509
T 114				-52		139	560
T 115				108		309	560
T 116				-43		130	560
T 117				128		284	560
T 118						385	377
T 119						>350	435
T 120						>350	435
T 121			$(1.33)^{23}$			249	434
T 122				-51		100	560, 3
T 123						64	120
T 124						364	377
T 125				-40		140	560
T 126				113		303	560, (377)

No.	Polymer	$\dfrac{\delta}{cal^{1/2}\,cm^{-3/2}}$	$\dfrac{\Delta H_u}{cal\,mol^{-1}}$	n
T 127	**Thietane, 1,1-dioxide,** polysulfone			
T 128	—, **3,3-pentamethylene-, 1,1-dioxide.** See **Thiaspiro**[3.5]**nonane, 2,2-dioxide.**			
T 129	**Thiocarbonyl fluoride.** See **Formaldehyde, difluorothio-.**			
T 130	**Thiophene, tetrahydro-,** poly(thioether)			
T 131	—, **3,4,5-trichloro-2-vinyl-**			
T 132	—, **2-vinyl-**			1.6376_D^{20}
T 133	**2,5-Thiophenedipropionic acid,** polyanhydride			
T 134	**Toluene, _m_-allyl-,** (stereoregular)			
T 135	—, _o_-**allyl-,** (stereoregular)			
T 136	—, _p_-**allyl-,** (stereoregular)			
T 137	—, 2,5-ethylene polymer			
T 138	_o_-**Toluic acid,** vinyl ester			
T 139	_p_-**Toluic acid,** vinyl ester			
T 140	_s_-**Triazine, 2,4-dimethyl-6-vinyl-**			
T 141	**Tributylvinylphosphonium bromide**			
T 142	**Tridecanedioic acid,** polyamide with 1,10-decanediamine			
T 143	—, polyamide with 1,5-pentanediamine			
T 144	—, polyamide with 2,2'-_p_-phenylenebis(ethylamine)			
T 145	—, polyamide with 1,13-tridecanediamine			1.536_D^{24}
T 146	—, polyamide with _p_-xylene-α,α'-diamine			
T 147	**Tridecanoic acid, 13-amino-,** polyamide			

No.	$\dfrac{10^4\,dv/dT}{\text{cc g}^{-1}\,\text{deg}^{-1}}$	d_a g cm^{-3}	d_c g cm^{-3}	T_g °C	T_d °C	T_m °C	References
T 127						300	377, (560)
T 128							
T 129							
T 130						67	120
T 131		1.69^{33}					304
T 132							311
T 133						78	229
T 134		(1.037)		40	75_{100}	180	147
T 135		(1.031)		80	115_{100}	290	147
T 136		(1.035)		65	105_{100}	240	147
T 137					55	230	247
T 138				44			93
T 139					-179_{9950}		556
T 140				77			93
T 141				16			356
T 142						175	352
T 143						176	352
T 144			$(1.095)^{25}$			262	529
T 145			(1.024)			172	561
T 146			$(1.108)^{25}$		100_{110}	247	529
T 147	(4.4)/(5.2)		$(1.018)^{30}$	41		183	106

No.	Polymer	$\dfrac{\delta}{cal^{1/2}\ cm^{-3/2}}$	$\dfrac{\Delta H_u}{cal\ mol^{-1}}$	n
U 1	**Undecanedioic acid,** polyamide with 1,4-butanediamine			
U 2	−, polyamide with 1,5-pentanediamine			
U 3	−, polyamide with 2,2′-p-phenylenebis(ethylamine)			
U 4	−, polyamide with p-xylene-α,α'-diamine			
U 5	**Undecanoic acid, 11-amino-,** polyamide		9900	
U 6	−, **11-(4-carboxybutoxy)-,** polyamide with 1,6-hexanediamine			
U 7	−, −, polyamide with 5,5′-oxybis(pentylamine)			
U 8	−, −, polyamide with p-xylene-α,α'-diamine			
U 9	−, **11-(p-carboxyphenoxy)-,** polyester with ethylene glycol			
U 10	−, −, polyester with 1,6-hexanediol			
U 11	−, **11-(3-carboxypropoxy)-,** polyamide with 1,6-hexanediamine			
U 12	−, −, polyamide with p-xylene-α,α'-diamine			
U 13	−, **11-methylamino-,** polyamide			
U 14	**1-Undecene, 11-iodo-**			
V 1	**Valeraldehyde,** polyacetal (isotactic)			
V 2	**Valeric acid, 5-amino-,** polyamide			
V 3	−, **5-(3-carboxypropoxy)-,** polyamide with 1,6-hexanediamine			
V 4	−, −, polyamide with 5,5′-oxybis(pentylamine)			
V 5	−, −, polyamide with p-xylene-α,α'-diamine			
V 6	−, **5-hydroxy-,** polyester			1.465_D^{50}
V 7	−, **nonafluoro-,** vinyl ester			1.353_D^{25}
V 8	−, **5,5′-oxydi-,** polyamide with 1,6-hexanediamine			
V 9	−, −, polyamide with 5,5′-oxybis(pentylamine)			
V 10	−, −, polyamide with p-xylene-α,α'-diamine			
V 11	−, **5,5′-thiodi-,** polyamide with 1,6-hexanediamine			

156

No.	$\dfrac{10^4 \, dv/dT}{\text{cc g}^{-1}\,\text{deg}^{-1}}$	$\dfrac{d_a}{\text{g cm}^{-3}}$	$\dfrac{d_c}{\text{g cm}^{-3}}$	$\dfrac{T_g}{°C}$	$\dfrac{T_d}{°C}$	$\dfrac{T_m}{°C}$	References
U 1						208	352
U 2						176	157, (352)
U 3			$(1.108)^{25}$		96_{110}	275	529
U 4			$(1.130)^{25}$	107	110_{110}	264	529
U 5	$(3.6)/(4.5)$	$(1.0255)^{25}$	1.228	46	$56_{3.8}$	198	570, 106, 162, 195, 16
					$-55_{6.8}$		16, (321), (348), (350)
					$-115_{8.3}$		16
U 6						160	527
U 7						128	527
U 8						223	527
U 9						65	55
U 10						72	55
U 11						159	527
U 12						220	527
U 13			$(1.005)^{30}$	-12		80	106
U 14					-6_{210}		486
					-120_{390}		486
V 1						155 I	405
						85 II	405
V 2		(1.240)				259	440
V 3						160	527
V 4						127	527
V 5						234	527
V 6						<50	490
V 7				20			359
V 8						180	527
V 9						134	527
V 10						243	527
V 11						185	530

No.	Polymer	$\dfrac{\delta}{\text{cal}^{1/2}\,\text{cm}^{-3/2}}$	$\dfrac{\varDelta H_u}{\text{cal mol}^{-1}}$	n
V 12	**Valeric acid, 5,5′-thiodi-,** polyamide with p-xylene-α,α'-diamine			
V 13	**Vanillin,** polymercaptal with 1,6-hexanedithiol			
V 14	**Vinyl acetate**	9.35		1.467_D^{20}
				1.4662_D^{30}
V 15	**Vinyl alcohol**		1640	1.5_D^{20}
V 16	**Vinyl bromide.** See **Ethylene, bromo-.**			
V 17	**Vinyl chloride.** See **Ethylene, chloro-.**			
V 18	**Vinyl fluoride.** See **Ethylene, fluoro-.**			
V 19	**Vinylidene bromide.** See **Ethylene, 1,1-dibromo-.**			
V 20	**Vinylidene chloride.** See **Ethylene, 1,1-dichloro-.**			
V 21	**Vinylidene fluoride.** See **Ethylene, 1,1-difluoro-.**			
V 22	**Vinyl butyrate.** See **Butyric acid,** vinyl ester.			
V 23	**Vinyl propionate.** See **Propionic acid,** vinyl ester.			
X 1	m-**Xylene, 5-allyl-**			
X 2	o-**Xylene, 4-allyl-**			
X 3	p-**Xylene, 2-allyl-**			
X 4	**2,6-Xylenol,** 1,4-polyether			
X 5	**Xylylene.** See **Benzene,** ethylene polymer *(alt)*.			

No.	$\dfrac{10^4 \text{ dv/dT}}{\text{cc g}^{-1}\text{ deg}^{-1}}$	$\dfrac{d_a}{\text{g cm}^{-3}}$	$\dfrac{d_c}{\text{g cm}^{-3}}$	$\dfrac{T_g}{°C}$	$\dfrac{T_d}{°C}$	$\dfrac{T_m}{°C}$	References
V 12						242	530
V 13						40	241
V 14	2.07/5.98	1.191^{20}		30	$46_{0.8}$		467, 326, 2, 199, 30, 565
	1.8/5.5	1.184^{28}			-30_{11}		190, 16, 368, 191
		1.181^{33}			-100_{12}		190, 16
		1.16^{60}					199, (450)
		1.13^{100}					199
V 15	3/	1.269	1.34	99		258	496, 326, 33, 284, 443
			1.345			133	90, 508, (385)
			1.35				442
V 16							
V 17							
V 18							
V 19							
V 20							
V 21							
V 22							
V 23							
X 1						252	147
X 2						275	147
X 3						338	147
X 4						261	238
X 5							

References

1. GORDON, M.: Structure and Physical Properties. London: Iliffe Books, Ltd., 1963.
2. TOBOLSKY, A. V.: Properties and Structure of Polymers. New York: John Wiley & Sons, Inc., 1960.
3. FLORY, P. J.: Principles of Polymer Chemistry. Ithaca: Cornell University Press, 1953.
4. MANDELKERN, L., and P. J. FLORY: J. Am. Chem. Soc. *73*, 3206 (1951).
5. HAMMER, C. F., T. A. KOCH, and I. F. WHITNEY: J. Appl. Polymer Sci. *1*, 169 (1959).
6. OVERBERGER, C. G., and E. B. DAVIDSON: J. Polymer Sci. *62*, 23 (1962).
7. SMALL, P. A.: J. Appl. Chem. *3*, 71 (1953).
8. RICHARDS, R. B.: Trans. Faraday Soc. *42*, 10 (1946).
9. SCOTT, R. L., and M. MAGAT: J. Polymer Sci. *4*, 555 (1949).
10. DOTY, P., and H. S. ZABLE: J. Polymer Sci. *1*, 90 (1946).
11. MAGAT, M.: J. Chim. Phys. *46*, 344 (1949).
12. BUNN, C. W.: J. Polymer Sci. *16*, 323 (1955).
13. GEE, G.: Trans. Faraday Soc. *38*, 418 (1942).
14. SPURLIN, H. M.: Ch. 10 in Cellulose and Cellulose Derivatives. New York: Interscience Publishers, Inc., 1955.
15. SIMRIL, V. L.: J. Polymer Sci. *2*, 147 (1947).
16. SCHMIEDER, K., and K. WOLF: Kolloid-Z. *134*, 149 (1953).
17. REHBERG, C. E., and C. H. FISHER: Ind. Eng. Chem. *40*, 1429 (1948).
18. BEEVERS, R. B., and E. F. T. WHITE: Trans. Faraday Soc. *56*, 1529 (1960).
19. DAVIES, T. E.: Brit. Plastics *32*, 283 (1959).
20. WILLBOURN, A. H.: Trans. Faraday Soc. *54*, 717 (1958).
21. DIMARZIO, E. A., and J. H. GIBBS: J. Polymer Sci *40*, 121 (1959).
22. DOLE, M.: Kolloid-Z. *165*, 40 (1959).
23. DOLE, M.: Fortschr. Hochpolymer. Forsch. *2*, 221 (1960).
24. HODES, W.: American Cyanamid Co. Unpublished Results.
25. KELL, R. M., B. BENNETT, and P. B. STRICKNEY: J. Appl. Polymer Sci. *2*, 8 (1959).
26. KRIGBAUM, W. R., and N. TOKITA: J. Polymer Sci. *43*, 467 (1960).
27. LINTON, W. H., and H. H. GOODMAN: J. Appl. Polymer Sci. *1*, 179 (1959).
28. MANARESI, P., and V. GIANNELLA: J. Appl. Polymer Sci. *4*, 251 (1960).
29. NEWMAN, S., and W. P. COX: J. Polymer Sci. *46*, 29 (1960).
30. SAITO, S., and T. NAKAJIMA: J. Appl. Polymer Sci. *2*, 93 (1959).
31. ROGERS, S. S., and L. MANDELKERN: J. Phys. Chem. *61*, 985 (1957).
32. KOLB, H. J., and E. F. IZARD: J. Appl. Phys. *20*, 564 (1949).
33. ROFF, W. J.: Fibers, Plastics and Rubbers. New York: Academic Press 1956.
34. BUNN, C. W.: J. Appl. Phys. *25*, 820 (1954).
35. BUECHE, A. M.: J. Am. Chem. Soc. *74*, 65 (1952).
36. FLORY, P. J., L. MANDELKERN, and H. K. HALL: J. Am. Chem. Soc. *73*, 2532 (1951).
37. MANDELKERN, L., and F. A. QUINN, Jr.: J. Appl. Phys. *25*, 830 (1954).
38. NICHOLS, J. B.: J. Appl. Phys. *25*, 847 (1954).
39. OVERBERGER, C. G., C. FRAZIER, J. MANDELMAN, and H. F. SMITH: J. Am. Chem. Soc. *75*, 3326 (1953).
40. SAUER, J. A., A. E. WOODWARD, and N. FUSCHILLO: J. Appl. Phys. *30*, 1488 (1959).
41. SAUER, J. A., R. A. WALL, N. FUSCHILLO, and A. E. WOODWARD: J. Appl. Phys. *29*, 1385 (1958).
42. FITZGERALD, E. R.: J. Appl. Phys. *29*, 1442 (1958).
43. IMOTO, M., and I. SOEMATSU: Bull. Chem. Soc. Japan *34*, 26 (1961).
44. NATTA, G., L. PORRI, P. CORRADINI, G. ZANINE, and F. CIAMPELLA: Atti Accad. Nazl. Lincei, Rend. Classe Sci. Fis. Mat. Nati. *29*, 257 (1960).
45. ISHIDA, Y., M. MATSUO, H. ITO, M. YOSHINO, F. IRIE, and M. TAKAYANAGI: Kolloid-Z. *174*, 162 (1961).
46. REDING, F. P.: J. Polymer Sci. *21*, 547 (1956).
47. NATTA, G., F. DANUSSO, and G. MORAGLIO: J. Polymer Sci. *25*, 119 (1957).

48. McCRUM, N. G.: J. Polymer Sci. *34*, 355 (1959).
49. FURAKOWA, G. T., R. E. McCOSKEY, and G. J. KING: J. Res. Natl. Bur. Std. *49*, 273 (1952).
50. BEEVERS, R. B., and E. F. T. WHITE: Trans. Faraday Soc. *56*, 744 (1960).
51. GRIEVESON, B. M.: Polymer *1*, 499 (1960).
52. BOVEY, F. A., J. F. ALBERE, G. B. RATHMANN, and C. L. SANDBERG: J. Polymer Sci. *15*, 520 (1955).
53. MILLER, M. L., and C. E. RAUHUT: J. Polymer Sci. *38*, 63 (1959).
54. KELL, R. M., B. BENNETT, and P. B. STRICKNEY: Rubber Chem. Technol. *31*, 499 (1958).
55. IZARD, E. F.: J. Polymer Sci. *8*, 503 (1952).
56. MARK, H., and A. V. TOBOLSKY: Physical Chemistry of High Polymeric Systems. New York: Interscience Publishers, Inc., 1950.
57. BILLMEYER, F. W. Jr.: Polymer Chemistry. New York: Interscience Publishers, Inc., 1957.
58. ROBERTS, D. E., and L. MANDELKERN: J. Am. Chem. Soc. *77*, 781 (1955).
59. MANDELKERN, L., F. A. QUINN, Jr., and D. E. ROBERTS: J. Am. Chem. Soc. *78*, 926 (1956).
60. NATTA, G., and P. CORRADINI: J. Polymer Sci. *20*, 251 (1956).
61. WHITBY, G. S.: Synthetic Rubber. New York: John Wiley & Sons, Inc., 1954.
62. WILEY, R. H., and G. M. BRAUER: J. Polymer Sci. *3*, 647 (1948).
63. VOGEL, O.: J. Polymer Sci. *46*, 261 (1960).
64. COLE, E. A., and D. R. HOLMES: J. Polymer Sci. *46*, 245 (1960).
65. GRIFFITH, J. H., and B. G. RÅNBY: J. Polymer Sci. *44*, 369 (1960).
66. MILLER, R. L., and L. E. NIELSEN: J. Polymer Sci. *44*, 391 (1960).
67. BACCAREDDA, M., and E. BUTTA: J. Polymer Sci. *44*, 421 (1960).
68. OVERBERGER, C. G., A. E. BORCHERT, and A. KATCHMAN: J. Polymer Sci. *44*, 491 (1960).
69. WOODWARD, A. E., J. M. CRISSMAN, and J. A. SAUER: J. Polymer Sci. *44*, 23 (1960).
70. LENZ, R. W., and C. E. HANDLOVITS: J. Polymer Sci. *43*, 167 (1960).
71. NIELSEN, L. E.: J. Polymer Sci. *42*, 357 (1960).
72. SWAN, P. R.: J. Polymer Sci. *42*, 525 (1960).
73. GOLIKE, R. C.: J. Polymer Sci. *42*, 583 (1960).
74. KE, B.: J. Polymer Sci. *42*, 15 (1960).
75. YU, A. J., and R. D. EVANS: J. Polymer Sci. *42*, 249 (1960).
76. GAYLORD, H. G., and H. F. MARK: Linear and Stereoregular Addition Polymers. New York: Interscience Publishers, Inc., 1959.
77. DUNHAM, K. R., J. W. H. FABER, J. VANDENBERGHE, and W. F. FOWLER, Jr.: J. Appl. Polymer Sci. *7*, 897 (1963).
78. VANDENBERG, E. J.: J. Polymer Sci. *47*, 486 (1960).
79. FARROW, G. J. McINTOSH, and I. M. WARD: Makromol. Chem. *38*, 147 (1960).
80. NATTA, G., G. MAZZANTI, P. CORRADINI, P. CHINI, and I. W. BASSI: Atti Accad. Nazl. Lincei, Rend. Classe Sci. Fis. Mat. Nati. *28*, 8 (1960).
81. McCULLOUGH, J. D., R. S. BAUER, and T. L. JACOBS: Chem. Ind. (London), 706 (1957).
82. YIN, T. P., and J. D. FERRY: J. Colloid Sci. *16*, 166 (1961).
83. ANAGNOSTOPOULOS, C. E., A. Y. CORAN, and H. R. GAMRATH: J. Appl. Polymer Sci. *4*, 181 (1960).
84. CANALE, A. J., J. B. KINSINGER, J. R. PANCHAK, R. L. KELSO, and R. K. GRAHAM: J. Appl. Polymer Sci. *4*, 231 (1960).
85. KORSHAK, V. V., T. M. FRUNZE, and L. V. KOSLOV: Vysokomolekul. Soedin. *2*, 838 (1960).
86. FIELD, N. D.: J. Polymer Sci. *47*, 518 (1960).
87. FOX, T. G., B. S. GARRETT, W. E. GOODE, S. GRATCH, J. F. KINCAID, A. SPELL, and J. D. STROUPE: J. Am. Chem. Soc. *80*, 1768 (1958).
88. MANDELKERN, L., G. M. MARTIN, and F. A. QUINN, Jr.: J. Res. Natl. Bur. Std. *58*, 137 (1957).
89. SCHILDKNECHT, C. E., S. T. GROSS, H. R. DAVIDSON, J. M. LAMBERT, and A. O. ZOSS: Ind. Eng. Chem. *40*, 2104 (1948).

162

90. HUGHES, L. H., and D. B. FORDYCE: J. Polymer Sci. *22*, 509 (1956).
91. OHLBERG, S. M., and S. S. FENSTERMAKER: J. Polymer Sci. *32*, 514 (1958).
92. WEIR, C. E., W. H. LESER, and L. A. WOOD: J. Res. Natl. Bur. Std. *44*, 367 (1950).
93. EBERLIN, E. C.: American Cyanamid Co. Unpublished Results.
94. WOOTEN, W. C., and H. W. COOVER, Jr.: J. Polymer Sci. *37*, 560 (1959).
95. REDING, F. P., and E. R. WALTER: J. Polymer Sci. *37*, 555 (1959).
96. BARB, W. G.: J. Polymer Sci. *37*, 515 (1959).
97. CLARK, K. J., A. TURNER JONES, and D. J. H. SANDIFORD: Chem. Ind. (London), 2010, 1962.
98. GRAHAM, P. J., E. L. BUHLE, and N. PAPPAS: J. Org. Chem. *26*, 4658 (1961).
99. SCHILLER, A. M., J. C. PETROPOULOS, and C. S. H. CHEN: J. Appl. Polymer Sci. *8*, 1699 (1964).
100. HATANO, M., and S. KAMBARA: Polymer *2*, 1 (1961).
101. YIN, T. P., S. E. LOVELL, and J. D. FERRY: J. Phys. Chem. *65*, 534 (1961).
102. IWAKURA, Y., Y. TANEDA, and S. UCHIDA: J. Appl. Polymer Sci. *5*, 108 (1961).
103. STARKWEATHER, H. W., and R. H. BOYD: J. Phys. Chem. *64*, 410 (1960).
104. NATTA, G., G. MAZZANTI, P. CORRADINI, A. VALVASSORI, and I. W. BASSI: Atti Accad. Nazl. Lincei, Rend. Classe Sci. Fis. Mat. Nat. *28*, 18 (1960).
105. ISHIDA, S.: Bull. Chem. Soc. Japan *33*, 727 (1960).
106. CHAMPETIER, G., and J. P. PIED: Makromol. Chem. *44*, 64 (1961).
107. NATTA, G., G. MAZZANTI, G. F. PREGAGLIA, and M. BINAGHI: Makromol. Chem. *44*, 537 (1961).
108. NISHIOKA, A., H. WATANABE, K. ABE, and Y. SONO: J. Polymer Sci. *48*, 241 (1960).
109. BUTLER, K., P. R. THOMAS, and G. J. TYLER: J. Polymer Sci. *48*, 357 (1960).
110. FORTUNE, L. R., and G. N. MALCOLM: J. Phys. Chem. *64*, 934 (1960).
111. TOPCHIEV, A. V., E. A. MUSHINA, A. I. PERELMAN, and B. A. KRENTSEL: Dokl. Akad. Nauk SSR *130*, 344 (1966).
112. TAKEDA, M., K. TANAKA, and R. NAGAO: J. Polymer Sci. *57*, 517 (1962).
113. REDING, F. P., J. A. FAUCHER, and R. D. WHITMAN: J. Polymer Sci. *57*, 483 (1962).
114. WILKINSON, R. W., and M. DOLE: J. Polymer Sci. *58*, 1089 (1962).
115. CONIX, A.: Intern. Symp. Macromol. Chem. Montreal, *D39* (1961).
116. VOGELSONG, D. C.: J. Polymer Sci. *57*, 895 (1962).
117. CAMPBELL, T. W., and A. C. HAVEN, Jr.: J. Appl. Polymer Sci. *1*, 73 (1959).
118. ALSUP, R. G., J. O. PUNDERSON, and G. F. LEVERETT: J. Appl. Polymer Sci. *1*, 185 (1959).
119. CORRADINI, P., and P. GANIS: Nuovo Cimento, Suppl. *15*, Ser. 10, 104 (1960).
120. LAL, J., and G. S. TRICK: J. Polymer Sci. *50*, 13 (1961).
121. SHARPLES, A., and F. L. SWINTON: J. Polymer Sci. *50*, 87 (1961).
122. KE, B., and A. W. SISKO: J. Polymer Sci. *50*, 87 (1961).
123. BOYD, R. H.: J. Polymer Sci. *50*, 133 (1961).
124. BARNES, W. J., and F. P. PRICE: J. Polymer Sci. *50*, 525 (1961).
125. GARRETT, B. S., W. E. GOODE, S. GRATCH, J. F. KINCAID, C. L. LEVESQUE, A. SPELL, J. D. STROUPE, and W. II. WATANABE: J. Am. Chem. Soc. *81*, 1007 (1959).
126. IIAAS, H. C., E. S. EMERSON, and N. W. SCIIULER: J. Polymer Sci. *22*, 291 (1956).
127. SHIELDS, D. J., and H. W. COOVER, Jr.: J. Polymer Sci. *39*, 532 (1959).
128. COBBS, W. H., Jr., and R. L. BURTON: J. Polymer Sci. *10*, 275 (1953).
129. OVERBERGER, C. G., L. H. AROND, R. H. WILEY, and R. R. GARRETT: J. Polymer Sci. *7*, 431 (1951).
130. MANDELKERN, L.: Chem. Rev. *56*, 903 (1956).
131. HILL, R.: Chem. Ind. (London), 1083 (1954).
132. TROSSARELLI, L., E. CAMPI, and A. G. MASINI: Intern. Symp. Makromol. Weisbaden, III A 16 (1959).
133. MATSUSHITA, S., T. HIGASHIMURA, and S. OKAMURA: Kobunshi Kagaku *17*, 456 (1960).
134. NATTA, G., G. DALL'ASTA, G. MAZZANTI, I. PASQUON, A. VALVASSORI, and A. ZAMBELLI: J. Am. Chem. Soc. *83*, 3343 (1961).
135. NATTA, G., G. MAZZANTI, G. DALL'ASTA, and P. LONGI: Makromol. Chem. *37*, 160 (1960).
136. MOODY, F. B.: Am. Chem. Soc. Polymer Div. Preprints, *2*, No. 2, 285 1961.

137. SIMRIL, V. L., and B. A. CURRY: J. Appl. Polymer Sci. *4*, 62 (1960).
138. NATTA, G., P. CORRADINI, and I. W. BASSI: Nuovo Cimento, Suppl. *15*, Ser. 10, 68 (1960).
139. NATTA, G., P. CORRADINI, and I. W. BASSI: Nuovo Cimento, Suppl. *15*, Ser. 10, 83 (1960).
140. CORRADINI, P., and P. GANIS: Nuovo Cimento, Suppl. *15*, Ser. 10, 96 (1960).
141. TINYAKOVA, E. I., B. A. DOLGOPLOSK, R. N. KOVALEVSKAYA, and T. G. ZHURAVLEV: Dokl. Akad. Nauk SSSR *129*, 1306 (1959).
142. TRUETT, W. L., D. R. JOHNSON, I. M. ROBINSON, and B. A. MONTAGUE: J. Am. Chem. Soc *82*, 2337 (1960).
143. ISHIDA, S.: Bull. Chem. Soc. Japan *33*, 924 (1960).
144. GALL, W. G., and N. G. McCRUM: J. Polymer Sci. *50*, 489 (1961).
145. FOX, T. G., and S. LOSHAEK: J. Polymer Sci. *15*, 371 (1955).
146. JACOBS, H., and E. JENCKEL: Makromol. Chem. *47*, 72 (1961).
147. PRICE, J. A., M. R. LYTTON, and B. G. RANBY: J. Polymer Sci. *51*, 541 (1961).
148. PEPPEL, W. J.: J. Polymer Sci. *51*, S64 (1961).
149. EDWARDS, W. M., and I. M. ROBINSON (to E. I. DuPont de Nemours & CO.). U. S. 2, 710, 853 (1955).
150. BOENIG, H. V., N. WALKER, and E. H. MYERS: J. Appl. Polymer Sci. *5*, 384 (1961).
151. FISCHER, D. J.: J. Appl. Polymer Sci. *5*, 436 (1961).
152. HAYASHI, K., Y. KITANISHI, M. NISHII, and S. OKAMURA: Makromol. Chem. *47*, 237 (1961).
153. PELLON, J., and W. G. CARPENTER: J. Polymer Sci. Pt. A *1*, 863 (1963).
154. CONIX, A.: Ind. Eng. Chem. *51*, 147 (1959).
155. EDGAR, O. B., and R. HILL: J. Polymer Sci. *8*, 1 (1952).
156. SCHNELL, H.: Ind. Eng. Chem. *51*, 157 (1959).
157. HILL, R., and E. E. WALKER: J. Polymer Sci. *3*, 609 (1948).
158. TANAKA, K.: Bull. Chem. Soc. Japan *33*, 1133 (1960).
159. DALL'ASTA, G., and N. ODDO: Chim. Ind. (Milan) *42*, 1234 (1960).
160. PREGAGLIA, G., and M. BINAGHI: Gazz. Chim. Ital. *90*, 1554 (1960).
161. SPERATI, C. A., and H. W. STARKWEATHER, Jr.: Fortschr. Hochpolymer. Forsch. *2*, 465 (1961).
162. KAWAGUCHI, T.: J. Appl. Polymer Sci. *2*, 56 (1959).
163. ILLERS, K. H., and E. JENCKEL: Kolloid-Z. *165*, 84 (1959).
164. BAILEY, W. J., and E. T. YATES: J. Org. Chem. *25*, 1800 (1960).
165. CHIANG, R., and P. J. FLORY: J. Am. Chem. Soc. *83*, 2857 (1961).
166. READ, B. E., and G. WILLIAMS: Polymer *2*, 239 (1961).
167. MOORE, W. R., and R. P. SHELDON: Polymer *2*, 315 (1961).
168. ALLEN, G., G. GEE, and G. J. WILSON: Polymer *1*, 456 (1960).
169. HALL, H. T.: J. Am. Chem. Soc. *74*, 68 (1952).
170. CHRISTOPHER, W. F., and D. W. FOX: Polycarbonates. New York: Reinhold Publishing Corp., 1962.
171. GUINOT, M.: Genie Chim. *85*, (3) 85 (1961).
172. BRISTOW, G. M., and W. F. WATSON: Trans. Faraday Soc. *54*, 1731 (1958).
173. MATHESON, M. S., E. E. AUER, E. B. BEVILACQUA, and E. J. HART: J. Am. Chem. Soc. *71*, 497 (1949).
174. WATSON, M. T.: SPE (Soc. Plastics Engrs.) J. *17*, 1083 (1961).
175. HOFF, E. A. W., D. W. ROBINSON, and A. H. WILBOURN: J. Polymer Sci. *18*, 161 (1955).
176. CRAWFORD, J. W. C.: J. Soc. Chem. Ind. (London) *68*, 201 (1949).
177. WALKER, E. E.: J. Appl. Chem. (London) *2*, 470 (1952).
178. MERKER, R. L., and M. J. SCOTT: J. Polymer Sci. Pt. A *2*, 15 (1964).
179. DALL'ASTA, G., and I. W. BASSI: Chim. Ind. (Milan) *43*, 999 (1961).
180. WILFONG, R. E.: J. Polymer Sci. *54*, 385 (1961).
181. NOSHAY, A., and C. C. PRICE: J. Polymer Sci. *54*, 533 (1961).
182. BURRELL, H.: Interchem. Rev. *14*, 3, 31 (1955).
183. FISHBEIN, L., and B. F. CROWE: Makromol. Chem. *48*, 221 (1961).
184. POLMANTEER, K. E., and M. J. HUNTER: J. Appl. Polymer Sci. *1*, 3 (1959).
185. BOYE, C. A.: J. Polymer Sci. *55*, 275 (1961).

186. BEKKEDAHL, N.: J. Res. Natl. Bur. Std. *13*, 411 (1934).
187. MANDELKERN, L., M. TRYON, and F. A. QUINN, Jr.: J. Polymer Sci. *19*, 77 (1956).
188. LOSHAEK, S.: J. Polymer Sci. *15*, 391 (1955).
189. FERRY, J. D., and G. S. PARKS: J. Chem. Phys. *4*, 70 (1936).
190. WILEY, R. H., and G. M. BRAUER: J. Polymer Sci. *4*, 351 (1949).
191. GORDON, M., and J. S. TAYLOR: Rubber Chem. Technol. *26*, 323 (1953).
192. MASON, P.: J. Chem. Phys. *35*, 1523 (1961).
193. NATTA, G., I. PASQUON, P. CORRADINI, M. PERALDO, M. PEGORARO, and A. ZAMBELLI: Atti Accad. Nazl. Lincei, Rend. Classe Sci. Fis. Mat. Nati. *28*, 539 (1960).
194. NATTA, G., D. SIANESI, D. MORERO, I. W. BASSI, and G. CAPORICCIO: Atti Accad. Nazl. Lincei, Rend. Classe Sci. Fis. Mat. Nati. *28*, 552 (1960).
195. MILLER, R. L., and L. E. NIELSEN: J. Polymer Sci. *55*, 643 (1961).
196. CHAPIN, E. C., J. G. ABRAMO, and V. L. LYONS: J. Org. Chem. *27*, 2595 (1962).
197. ILLERS, K.: Z. Elektrochem. *65*, 679 (1961).
198. REDING, F. P., J. A. FAUCHER, and R. D. WHITMAN: J. Polymer Sci. *54*, S56 (1961).
199. SCHILDKNECHT, C. E.: Vinyl and Related Polymers. New York: John Wiley & Sons Inc., 1952.
200. OVERBERGER, C. G., and A. M. SCHILLER: J. Polymer Sci. *54*, S30 (1961).
201. Austr. Pat. Appl. 58950 (to Brit. Nylon Spinners, Ltd), March 29, 1960.
202. BIRNBOIM, M. H., and J. D. FERRY: J. Appl. Phys. *32*, 2305 (1961).
203. CHATANI, Y., T. TAKIYAWA, S. MURAHASHI, Y. SAKATA, and Y. NISHIMURA: J. Polymer Sci. *55*, 811 (1961).
204. BOHN, C. R., J. R. SCHAEFGEN, and W. O. STATTON: J. Polymer Sci. *55*, 531 (1961).
205. HILDEBRAND, J. H., and R. L. SCOTT: The Solubility of Nonelectrolytes. New York: Reinhold Publishing Corp., 1950.
206. SPENCER, R. S., and G. D. GILMORE: J. Appl. Phys. *20*, 502 (1949).
207. SHASHOUA, V. E., and W. M. EARECKSON: J. Polymer Sci. *40*, 343 (1959).
208. WILLIAMS, M. L., R. F. LANDEL, and J. D. FERRY: J. Am. Chem. Soc. *77*, 3701 (1955).
209. SMITH, T. L.: J. Polymer Sci. *32*, 99 (1958).
210. SCHAEFGEN, J. R., F. H. KOONTZ, and R. F. TIETZ: J. Polymer Sci. *40*, 377 (1959).
211. KELLEY, F. N., and F. BUECHE: J. Polymer Sci. *50*, 549 (1961).
212. WOOD, L. A.: J. Polymer Sci. *28*, 319 (1958).
213. CLARK, E. S., and L. T. MUUS: Z. Krist. *117*, 119 (1962).
214. HAYES, R. A.: J. Appl. Polymer Sci. *5*, 318 (1961).
215. MANDELKERN, L.: J. Polymer Sci. *47*, 494 (1960).
216. BOYER, R. F.: J. Appl. Phys. *25*, 825 (1954).
217. MAKIMOTO, T., T. TSURUTA, and J. FURUKAWA: Makromol. Chem. *50*, 116 (1961).
218. FURUKAWA, J., T. SAEGUSA, T. TSURUTA, S. OHTA, and G. WASAI: Makromol. Chem. *52*, 230 (1962).
219. FUJIO, R., T. TSURUTA, and J. FURUKAWA: Makromol. Chem. *52*, 233 (1962).
220. MANDELKERN, L.: J. Appl. Phys. *26*, 443 (1955).
221. KENNEDY, J. P., and R. M. THOMAS: Makromol. Chem. *53*, 28 (1962).
222. NATTA, G., L. PORRI, A. CARBONARO, and G. LUGLI: Makromol. Chem. *53*, 52 (1962).
223. CANALE, A. J., W. A. HEWETT, T. M. SHRYNE, and E. A. YOUNGMAN: Chem. Ind. (London), 1054 (1962).
224. GAWLAK, M., R. P. PALMER, J. B. ROSE, D. J. H. SANDIFORD, and A. TURNER-JONES: Chem. Ind (London), 1148 (1962).
225. RÅNBY, B. G., K. S. CHAN, and H. BRUMBERGER: J. Polymer Sci. *58*, 545 (1962).
226. HECK, R. F. (to Hercules Powder Co.): U.S. 3,048, 573 (1962).
227. GARFIELD, L. J., S. E. PETRIE, and D. W. VANAS: Trans. Soc. Rheol. *6*, 131 (1962).
228. NATTA, G., G. DALL'ASTA, G. MAZZANTI, I. PASQUON, A. VALVASSORI, and A. ZANBELLI: Makromol. Chem. *54*, 95 (1962).
229. YODA, N.: Preprints Am. Chem. Soc. Div. Polymer Chem. *3*, No. 2, 184 (1962).
230. JACKSON, W. J., Jr., and J. R. CALDWELL: J. Appl. Polymer Sci. *7*, 1975 (1963).
231. HOFFMAN, J. D., and J. J. WEEKS: J. Res. Natl. Bur. Std. *60*, 465 (1958).
232. ERREDE, L. A., and N. KNOLL: J. Polymer Sci. *60*, 33 (1962).
233. NATTA, G., G. MAZZANTI, G. F. PREGAGLIA, and G. POZZI: J. Polymer Sci. *58*, 1201 (1962).

234. HUGHES, L. J., and G. E. BRITT: J. Appl. Polymer Sci. *5*, 337 (1961).
235. Bulletin SP-251 (Rohm & Haas Co., Sept. 1962).
236. CALDWELL, J. R., and R. GILKEY (to Eastman Kodak Co).: U.S. 3,053,805 (1962).
237. DAINTON, F. S., D. M. EVANS, F. E. HOARE, and T. P. MELIA: Polymer *3*, 263 (1962).
238. BUTTE, W. A., C. C. PRICE, and R. E. HUGHES: J. Polymer Sci. *61*, S28 (1962).
239. YODA, N.: Makromol. Chem. *56*, 36 (1962).
240. NATTA, G., J. DiPIETRO, and M. CAMBINI: Makromol. Chem. *56*, 200 (1962).
241. GAYLORD, N. G.: Polyethers. III Polyalkylene Sulfides and Other Polythioethers. New York: Interscience Publishing Co., 1962.
242. NOSHAY, A., and C. C. PRICE: J. Polymer Sci. *34*, 165 (1959).
243. FLORY, P. J.: J. Am. Chem. Soc. *62*, 1057 (1940).
244. VANDENBERG, E. J., R. F. HECK, and D. S. BRESLOW: J. Polymer Sci. *41*, 519 (1959).
245. HECK, R. F., and D. S. BRESLOW: J. Polymer Sci. *41*, 520 (1959).
246. HECK, R. F., and D. S. BRESLOW: J. Polymer Sci. *41*, 521 (1959).
247. AUSPOS, L. A., C. W. BURNAM, L. HALL, K. J. HUBBARD, W. KIRK, J. R. SCHAEFGEN, and S. B. SPECK: J. Polymer Sci. *15*, 19 (1955).
248. INOUE, M.: J. Polymer Sci. *61*, 343 (1962).
249. MARK, H. F., S. M. ATLAS, and N. OGATA: J. Polymer Sci. *61*, S49 (1962).
250. WILLIAMS, J. L. R., T. M. LAAKSO, and L. E. CONTOIS: J. Polymer Sci. *61*, 353 (1962).
251. CONIX, A., and R. VAN KERPEL: J. Polymer Sci. *40*, 521 (1959).
252. CRAMER, F. B., and R. G. BEAMAN: J. Polymer Sci. *21*, 237 (1956).
253. BOCK, L. H., and J. K. ANDERSON: J. Polymer Sci. *17*, 553 (1955).
254. BOYER, R. F., and R. S. SPENCER: J. Appl. Phys. *15*, 398 (1944).
255. SAUNDERS, P. R., and J. D. FERRY: J. Colloid Sci. *14*, 239 (1959).
256. HEIJBOER, I. J.: Kolloid-Z. *148*, 36 (1956).
257. HEIJBOER, I. J.: Kolloid-Z. *171*, 7 (1960).
258. IZARD, E. J.: J. Polymer Sci. *9*, 35 (1952).
259. KAMBARA, S., and A. TAKAHASHI: Makromol. Chem. *58*, 226 (1962).
260. IIMURA, K., and S. KONDO: J. Appl. Polymer Sci. *6*, S52 (1962).
261. CHARLESBY, A., and L. CALLAGHAN: Phys. Chem. Solids *4*, 227 (1958).
262. SANDIFORD, D. J. H.: J. Appl. Chem. (London) *8*, 188 (1958).
263. THOMPSON, A. B., and D. W. WOODS: Trans. Faraday Soc. *52*, 1383 (1956).
264. HAUSER, R. L., C. A. WALKER, and F. L. KILBOURNE: Ind. Eng. Chem. *48*, 1202 (1956).
265. READ, B. E.: Polymer *3*, No. 4, 529 (1962).
266. STANLEY, E., and M. LITT: J. Polymer Sci. *43*, 453 (1960).
267. SMITH, K. L., and R. VAN CLEVE: Ind. Eng. Chem. *50*, 12 (1958).
268. CORRADINI, P., and P. GANIS: J. Polymer Sci. *43*, 311 (1960).
269. ILLERS, K. H., and H. BREUER: J. Colloid Sci. *18*, 1 (1963).
270. FARROW, G., and I. M. WARD: Polymer *1*, 330 (1960).
271. WATANABE, H., R. KOYAMA, H. NAGAI, and A. NISHIOKA: J. Polymer Sci. *62*, S74 (1962).
272. ZUTTY, N. L., F. P. REDING, and J. A. FAUCHER: J. Polymer Sci. *62*, S171 (1962).
273. TEARE, P. W., and D. R. HOLMES: J. Polymer Sci. *24*, 496 (1957).
274. NATTA, G., P. CORRADINI, and I. W. BASSI: Makromol. Chem. *33*, 247 (1959).
275. ZANNETTI, R., P. MANARESI, and G. C. BUZZONI: Chim. Ind. (Milan) *43*, 735 (1961).
276. ASAHINA, M., and K. OKUDA: Kobunshi Kagaku *17*, 607 (1960).
277. KINOSHITA, Y.: Makromol. Chem. *33*, 21 (1959).
278. REINHARDT, R. C.: Ind. Eng. Chem. *35*, 422 (1943).
279. NARITA, S., and K. OKUDA: J. Polymer Sci. *38*, 270 (1959).
280. DANNIS, M. L.: J. Appl. Polymer Sci. *7*, 231 (1963).
281. YEN, T. F.: J. Polymer Sci. *38*, 272 (1959).
282. YEN, T. F.: J. Polymer Sci. *35*, 533 (1959).
283. BUNN, C. W.: Proc. Roy. Soc. (London), Ser. A*180*, 40 (1942).
284. SAKURADA, I., Y. NUKUSHINA, and Y. SONE: IUPAC Symp. Macromol. Chem., 715 (1954).

285. BAPTIST, J. N., and F. X. WERBER: Chem. Eng. News *40* (1963).
286. PASSAGLIA, E., and K. H. KEVORKIAN: J. Appl. Phys. *34,* 90 (1963).
287. DANUSSO, F., G. MORAGLIO, W. GHIGLIA, L. MOTTA, and G. TALAMINI: Chim. Ind. (Milan) *41,* 748 (1959).
288. QUINN, F. A., Jr., and L. MANDELKERN: J. Am. Chem. Soc. *80,* 3178 (1958).
289. NATTA, G., L. PORRI, A. CARBONARO, F. CIAMPELLI, and G. ALLEGRA: Makromol. Chem. *51,* 229 (1962).
290. NATTA, G., L. PORRI, G. STOPPA, G. ALLEGRA, and F. CIAMPELLI: J. Polymer Sci. Pt. B *1,* 67 (1963).
291. BOOR, J., Jr., and J. C. MITCHELL: J. Polymer Sci. Pt. A *1,* 59 (1963).
292. DUNHAM, K. R., J. VANDENBERGHE, J. W. H. FABER, and L. E. CONTOIS: J. Polymer Sci. Pt A *1,* 751 (1963).
293. VANDENBERG, E. J.: J. Polymer Sci. Pt. C *1,* 207 (1963).
294. DANUSSO, F., and G. GIANOTTI: Makromol. Chem. *61,* 139 (1963).
295. DANUSSO, F., and G. POLIZZOTTI: Makromol. Chem. *61,* 157 (1963).
296. DANUSSO, F., and G. GIANOTTI: Makromol. Chem. *61,* 164 (1963).
297. PEREGO, G., and I. W. BASSI: Makromol. Chem. *61,* 198 (1963).
298. CARAZZOLO, G., and M. MAMMI: J. Polymer Sci. Pt. A *1,* 965 (1963).
299. VOGELSONG, D. C.: J. Polymer Sci. Pt. A *1,* 1055 (1963).
300. CRICK, F. H. C., and A. RICH: Nature *176,* 780 (1955).
301. BAMFORD, C. H., L. BROWN, E. M. CANT, A. ELLIOTT, W. E. HANBY, and B. R. MALCOLM: Nature *176,* 396 (1955).
302. COWAN, P. M., and S. McGAVIN: Nature *176,* 501 (1955).
303. WUNDERLICH, B., and M. DOLE: J. Polymer Sci. *32,* 125 (1958).
304. BACHMAN, G. B., L. J. FILAR, R. W. FINHOLT, L. V. HEISEY, H. M. HELLMAN, L. L. LEWIS, and D. D. MICUCCI: Ind. Eng. Chem. *43,* 997 (1951).
305. E. SMOLIN: American Cyanamid Co. Unpublished Results.
306. SHARPLES, A., and F. L. SWINTON: Polymer *4,* 119 (1963).
307. FLORY, P. J., H. D. BEDON, and E. H. KEEFER: J. Polymer Sci. *28,* 151 (1958).
308. WILHOIT, R. C., and M. DOLE: J. Phys. Chem. *57,* 14 (1953).
309. FURUYA, S., and M. HONDA: J. Polymer Sci. *28,* 232 (1958).
310. RUFFINO, G.: Paper 18 in Progress in International Research on Thermodynamic and Transport Properties. New York: Academic Press 1962.
311. Polaroid Corp., Off. Sci. Res. Dev. (OSRD) Report No. 4417, Feb. 1945.
312. HEWETT, W. A., and F. E. WEIR: J. Polymer Sci. Pt. A *1,* 1239 (1963).
313. CORRADINI, P., and P. GANIS: Makromol. Chem. *62,* 97 (1963).
314. SABA, R. G., J. A. SAUER, and A. E. WOODWARD: J. Polymer Sci. Pt. A *1,* 1483 (1963).
315. PRICE, C. C., M. OSGAN, R. E. HUGHES, and C. SHAMBELAN: J. Am. Chem. Soc. *78,* 690 (1956).
316. PIERRE, L. E. St., and C. C. PRICE: J. Am. Chem. Soc. *78,* 3432 (1956).
317. PRICE, C. C., and M. OSGAN: J. Am. Chem. Soc. *78,* 4787 (1956).
318. SHETTER, J. A.: J. Polymer Sci. Pt. B *1,* 209 (1963).
319. GIPSTEIN, E., E. WELLISH, and O. J. SWEETING: J. Polymer Sci. Pt. B *1,* 237 (1963).
320. DUNHAM, K. R., J. VANDENBERGHE, J. W. H. FABER, and W. F. FOWLER, Jr.: J. Appl. Polymer Sci. *7,* 143 (1963).
321. HORN, C. F., B. T. FREURE, H. VINEYARD, and H. J. DECKER: J. Appl. Polymer Sci. *7,* 887 (1963).
322. NATTA, G., L. PORRI, P. CORRADINI, and D. MORERO: Chim. Ind. (Milan) *40,* 362 (1958).
323. LITT, M.: J. Polymer Sci. Pt A *1,* 2219 (1963).
324. MITCHELL, J. C.: J. Polymer Sci. Pt. B *1,* 285 (1963).
325. GECHELE, G. B., and L. CRESCENTINI: J. Appl. Polymer Sci. *7,* 1349 (1963).
326. WURSTLIN, F.: Ch. 11 in Die Physik der Hochpolymeren, Vol. 3, H. A. Stuart, Ed. Berlin-Göttingen-Heidelberg: Springer 1955.
327. REHBERG, C. E., E. A. FAUCETTE, and C. H. FISHER: J. Am. Chem. Soc. *66,* 1723 (1944).
328. MANGARAJ, D., S. PATRA, and S. RASHID: Makromol. Chem. *65,* 29, 39 (1963).

167

329. LAWSON, K. D., J. A. SAUER, and A. E. WOODWARD: J. Appl. Phys. *34*, 2492 (1963).
330. WILLMOTT, P. W. T., and F. W. BILLMEYER, Jr.: Off. Dig. Federation Soc. Paint Technol. *35*, 847 (1963).
331. NATTA, G., and G. MORAGLIO: Makromol. Chem. *66*, 218 (1963).
332. SAKURADA, Y., K. IMAI, and M. MATSUMOTO: Kobunshi Kagaku *20*, 429 (1963).
333. HELLER, J., and D. J. LYMAN: J. Polymer Sci. Pt. B *1*, 317 (1963).
334. TURNER JONES, A., and J. M. AIZLEWOOD: J. Polymer Sci. Pt. B *1*, 471 (1963).
335. COOPER, W., and G. VAUGHAN: Polymer *4*, 329 (1963).
336. BECK, D. L., A. A. HILTZ, and J. R. KNOX: SPE (Soc. Plastics Engrs.) Trans. *3*, 279 (1963).
337. WASAI, G., T. TSURUTA, J. FURUKAWA, and R. FUJIO: Nippon Kagaku Zasshi *66*, 1339 (1963).
338. BANKS, W., and S. SHARPLES: Makromol. Chem. *67*, 42 (1963).
339. MANGARAJ, D., S. PATKA, and S. B. RATH: Makromol. Chem. *67*, 84 (1963).
340. ODA, R., T. SHONO, A. OKU, and H. TAKAO: Makromol. Chem. *67*, 124 (1963).
341. HALL, H. K., Jr.: J. Org. Chem. *28*, 3213 (1963).
342. REINKING, N. H., A. E. BARNABEO, and W. F. HALE: J. Appl. Polymer Sci. *7*, 2135 (1963).
343. NAKATSUKA, R.: Bull. Chem. Soc. Japan *36*, 1294 (1963).
344. GEACINTOV, C., R. S. SCHOTLAND, and R. B. MILES: J. Polymer Sci. Pt. B *1*, 587 (1963).
345. TURNER, A., and F. E. BAILEY, Jr.: J. Polymer Sci. Pt. B *1*, 587 (1963).
346. ROLDAN, L. G., and H. S. KAUFMAN: J. Polymer Sci. Pt. B *1*, 603 (1963).
347. WITTBECKER, E. L.: Chem. Eng. News *41* (1963).
348. PAUL, M., and E. SCHWARTZ: Festschrift Carl Wurster p. 333 (Badische Anilin- & Soda Fabrik AG, 1960).
349. ALLEN, G.: Techniques of Polymer Science, p. 167. S. C. I. Monograph No. 17. New York: Gordon Breach 1963.
350. COFFMAN, D. D., N. L. COX, E. L. MARTIN, W. E. MOCHEL, and F. J. VAN NATTA: J. Polymer Sci. *3*, 85 (1948).
351. WILEY, R. H., and G. M. BRAUER: J. Polymer Sci. *3*, 708 (1948).
352. COFFMAN, D. D., G. J. BERCHET, W. R. PETERSON, and E. W. SPANAGEL: J. Polymer Sci. *2*, 306 (1947).
353. PORT, W. S., J. E. HANSEN, E. F. JORDAN, Jr., T. T. DIETZ, and D. SWERN: J. Polymer Sci. *7*, 207 (1951).
354. CANNEPIN, A., G. CHAMPTIER, and A. PARISOT: J. Polymer Sci. *8*, 35 (1952).
355. ILYINA, D. E., B. A. KRENTSEL, and G. E. SEMENIDO: J. Polymer Sci. Pt. C *4*, Part 2, 999 (1963).
356. CHEN, C. S. H., and D. G. GRABAR: J. Polymer Sci. Pt. C *4*, 849 (1963).
357. TURNER JONES, A., and R. P. PALMER: Polymer *4*, 525 (1963).
358. DOAK, K. W., and H. N. CAMPBELL: J. Polymer Sci. *18*, 215 (1955).
359. REID, T. S., D. W. CODDING, and F. A. BOVEY: J. Polymer Sci. *18*, 417 (1955).
360. YODA, N., and I. MATSUBARA: J. Polymer Sci. Pt. A *2*, 253 (1964).
361. FOX, C. J.: J. Polymer Sci. Pt. A *2*, 267 (1964).
362. MORGAN, P. W.: J. Polymer Sci. A *2*, 437 (1964).
363. RUSSELL, J., and R. G. VAN KERPEL: J. Polymer Sci. *25*, 77 (1957).
364. GOODMAN, P.: J. Polymer Sci. *24*, 307 (1957).
365. EDWARDS, W. M., and I. M. ROBINSON (to E. I. Du Pont de Nemours & Co.): U. S. 2, 880, 230 (1959).
366. MARCONI, W., S. CESCA, and G. DELLA FORTUNA: J. Polymer Sci. Pt. B *2*, 301 (1964).
367. ALLEN, G.: J. Appl. Chem. (London) *14*, 1 (1964).
368. WILEY, R. H., and G. M. BRAUER: J. Polymer Sci. *11*, 221 (1953).
369. LEEPER, H. M., and W. SCHLESINGER: J. Polymer Sci. *11*, 307 (1953).
370. DULMADGE, W. J.: J. Polymer Sci. *26*, 277 (1957).
371. LEVINE, M., and S. C. TEMIN: J. Polymer Sci. *28*, 179 (1958).
372. TURNER JONES, A.: Makromol. Chem. *71*, 1 (1964).
373. McCURDY, R. M., and J. H. PRAGER: J. Polymer Sci. Pt A *2*, 1185 (1964).
374. OVERBERGER, C. G., and J. WEISE: J. Polymer Sci. Pt. B *2*, 329 (1964).

375. CESARI, M.: J. Polymer Sci. Pt. B *2*, 453 (1964).
376. TAKAHASHI, A., and S. KAMBARA: Makromol. Chem. *72*, 92 (1964).
377. FOLDI, V. S., and W. SWEENEY: Makromol. Chem. *72*, 208 (1964).
378. IWAKURA, Y., M. SAKAMOTO, and Y. AWATA: J. Polymer Sci. Pt. A *2*, 881 (1964).
379. CONIX, A.: J. Polymer Sci. *29*, 343 (1958).
380. PRAGER, J. H., R. M. McCURDY, and G. B. RATHMANN: J. Polymer Sci. Pt. A 2, 1941 (1964).
381. NATTA, G., G. MAZZANTI, P. LONGI, and F. BERNARDINI: J. Polymer Sci. *31*, 181 (1958).
382. NOEL, C.: Compt. Rend. *258*, 3702 (1964).
383. REDING, F. P., and E. R. WALTER: J. Polymer Sci. *37*, 555 (1959).
384. KIBLER, C. J., A. BELL, and J. G. SMITH: J. Polymer Sci. Pt. A *2*, 2115 (1964).
385. FUJII, K., T. MOCHIZUKI, S. IMOTO, J. UKIDA, and M. MATSUMOTO: J. Polymer Sci. Pt A *2*, 2327 (1964).
386. WILSKI, H., and T. GREWER: J. Polymer Sci. Pt. C *6*, 33 (1964).
387. O'REILLY, J. M., F. E. KARASZ, and H. E. BAIR: J. Polymer Sci. Pt. C *6*, 109 (1964).
388. GEACINTOV, C., R. S. SCHOTLAND, and R. B. MILES: J. Polymer Sci. Pt. C *6*, 197 (1964).
389. HIRONO, K., G. WASAI, T. SAEGUSA, and J. FURUKAWA: Kogyo Kagaku Zasshi *67*, 604 (1964).
390. KENNEDY, J. P., J. J. ELLIOTT, and B. GROTEN: Makromol. Chem. *77*, 26 (1964).
391. NATTA, G., L. PORRI, A. CARBONARO, and G. STOPPA: Makromol. Chem. *77*, 114 (1964).
392. INGRAHAM, F. S., and D. F. WOOLEY, Jr.: Ind. Eng. Chem. *56*, 53 (1964).
393. AUFDERMARSH, C. A., J., and R. PARISER: J. Polymer Sci. Pt. A *2*, 4727 (1964).
394. STRATTA, J. J., F. P. REDING, and J. A. FAUCHER: J. Polymer Sci. Pt. A *2*, 5017 (1964).
395. OHZAWA, Y., and Y. WADA: Japanese J. Appl. Phys. *3*, 436 (1964).
396. HARA, T., and S. OKAMOTO: Japanese J. Appl. Phys. *3*, 499 (1964).
397. KIRSHENBAUM, I., Z. W. WILCHINSKY, and B. GROTTEN: J. Appl. Polymer Sci. *8*, 2723 (1964).
398. ISAACSON, R. B., I. KIRSHENBAUM, and W. C. FEIST: J. Appl. Polymer Sci. *8*, 2789 (1964).
399. MARCONI, W., A. MAZZEI, S. CUCINELLA, and M. CESARI: J. Polymer Sci. Pt. A *2*, 4261 (1964).
400. ISHIBASHI, M.: J. Polymer Sci. Pt. A *2*, 4361 (1964).
401. KETLEY, A. D., and R. J. EHRIG: J. Polymer Sci. Pt. A *2*, 4461 (1964).
402. MILLER, M. L., and J. SKOGMAN: J. Polymer Sci. Pt. A *2*, 4551 (1964).
403. LAL, K., and G. S. TRICK: J. Polymer Sci. Pt. A *2*, 4559 (1964).
404. VOGL, O.: J. Polymer Sci. Pt. A *2*, 4591 (1964).
405. VOGL, O.: J. Polymer Sci. Pt. A *2*, 4621 (1964).
406. ALLEN, G., C. BOOTH, M. N. Jones, D. J. MARKS, and W. D. TAYLOR: Polymer *5*, 547 (1964).
407. CRISSMAN, J. M., J. A. SAUER, and A. E. WOODWARD: J. Polymer Sci. Pt. A *2*, 5075 (1964).
408. DANUSSO, F., and G. GIANOTTI: Makromol. Chem. *80*, 1 (1964).
409. TSURUTA, T., R. FUJIO, and J. FURUKAWA: Makromol. Chem. *80*, 172 (1964).
410. BELL, A., J. G. SMITH, and C. J. KIBLER: J. Polymer Sci. Pt. A *3*, 19 (1965).
411. ISHIDA, Y., H. ITO, and M. TAKAYANAGI: J. Polymer Sci. Pt. B *3*, 87 (1965).
412. KRIGBAUM, W. R., and I. UEMATSU: J. Polymer Sci. Pt. A *3*, 767 (1965).
413. MAGILL, J. H.: J. Polymer Sci. Pt. A *3*, 1195 (1965).
414. YAMADERA, R., and C. SONODA: J. Polymer Sci. Pt. B *3*, 411 (1965).
415. NATTA, G., P. CORRADINI, D. SIANESI, and D. MORERO: J. Polymer Sci. *51*, 527 (1961).
416. INOUE, M.: J. Polymer Sci. *51*, S18 (1961).
417. KIRSHENBAUM, I., R. B. ISAACSON, and M. DRUIN: J. Polymer Sci. Pt. B *3*, 525 (1965).
418. FATOU, J. G., C. H. BAKER, and L. MANDELKERN: Polymer *6*, 243 (1965).
419. TURNER JONES, A.: Polymer *6*, 249 (1965).

420. WITTBECKER, E. L., W. S. SPLIETHOFF, and C. R. STINE: J. Appl. Polymer Sci. *9*, 213 (1965).
421. EISENBERG, A., and L. TETER: J. Am. Chem. Soc. *87*, 2108 (1965).
422. SEDLAK, J. A., and K. MATSUDA: J. Polymer Sci. Pt. A. *3*, 2329 (1965).
423. DANUSSO, F., and G. GIANOTTI: J. Polymer Sci. Pt. B *3*, 537 (1965).
424. STARKWEATHER, H. W., Jr., and R. E. MOYNIHAN: J. Polymer Sci. *22*, 363 (1956).
425. HOLMES, D. R., C. W. BUNN, and D. J. SMITH: J. Polymer Sci. *17*, 159 (1955).
426. RICHARDS, J. R.: Dissertation Abstr. *22*, 1029 (1961).
427. BROWN, J. F., Jr., and D. W. WHITE: J. Am. Chem. Soc. *82*, 5671 (1960).
428. EVANS, R. D., H. R. MIGHTON, and P. J. FLORY: J. Am. Chem. Soc. *72*, 2018 (1950).
429. ARAKI, A.: J. Appl. Polymer Sci. *9*, 421 (1965).
430. KINSINGER, J. B., J. R. PANCHAK, R. L. KELSO, J. S. BARTLETT, and R. K. GRAHAM: J. Appl. Polymer Sci. *9*, 421 (1965).
431. MERCIER, J. P., J. J. AKLONIS, M. LITT, and A. V. TOBOLSKY: J. Appl. Polymer Sci. *9*, 447 (1965).
432. TEMIN, S. C.: J. Appl. Polymer Sci. *9*, 471 (1965).
433. PETERLIN, A., and J. D. HOLBROOK: Kolloid-Z. *203*, 68 (1965).
434. SHEEHAN, W. C., T. B. COLE, and L. G. PICKLESIMER: J. Appl. Polymer Sci. *9*, 1455 (1965).
435. MULVANEY, J. E., and C. S. MARVEL: J. Org. Chem. *26*, 95 (1961).
436. OTSUKA, S., K. MORI, and F. IMAIZUMI: J. Am. Chem. Soc. *87*, 3017 (1965).
437. MIKI, K., and R. NAKATSUKA: Rept. Progr. Polymer Phys. Japan *8*, 115 (1965).
438. WOODWARD, A. E., J. A. SAUER, and R. A. WALL: J. Polymer Sci. *50*, 117 (1961).
439. SAPPER, D. I.: J. Polymer Sci. *43*, 383 (1960).
440. YODA, N., and A. MIYAKE: J. Polymer Sci. *43*, 117 (1960).
441. BAKER, W. P., Jr.: J. Polymer Sci. Pt. A *1*, 655 (1963).
442. BUNN, C. W.: Nature *161*, 929 (1948).
443. TADOKORO, H., S. SEKI, and I. NITTA: Bull. Chem. Soc. Japan *28*, 559 (1955).
444. NATTA, G.: Makromol. Chem. *35—36*, 94 (1960).
445. WUNDERLICH, B., and D. POLAND: J. Polymer Sci. *58*, 1106 (1962).
446. NATTA, G., F. DANUSSO, and D. SIANESI: Makromol. Chem. *28*, 253 (1958).
447. SIANESI, D., and G. CAPORICCIO: Makromol. Chem. *60*, 213 (1963).
448. NATTA, G., G. DALL'ASTA, G. MAZZANTI, and A. CASALE: Makromol. Chem. *58*, 217 (1962).
449. NASINI, A. G., L. TROSSARELLI, and G. SAINI: Makromol. Chem. *44—46*, 550 (1961).
450. MICHAILOW, G. P.: Makromol. Chem. *35—36*, 26 (1960).
451. POWERS, J., J. D. HOFFMAN, J. J. WEEKS, and F. A. QUINN, Jr.: J. Res. Natl. Bur. Std. A. *69*, 335 (1965).
452. CRISSMAN, J. M., A. E. WOODWARD, and J. A. SAUER: J. Polymer Sci. Pt. A *3*, 2693 (1965).
453. INGHAM, J. D., and D. D. LAWSON: J. Polymer Sci. Pt. A *3*, 2707 (1965).
454. UNISHI, T., and M. HASEGAWA: J. Polymer Sci. Pt. A *3*, 3191 (1965).
455. STUEBEN, K. C.: J. Polymer Sci. Pt. A *3*, 3209 (1965).
456. TADOKORO, H., Y. TAKAHASHI, S. OTSUKA, K. MORI, and F. IMAIZUMI: J. Polymer Sci. Pt. B *3*, 697 (1965).
457. KENNEDY, J. P., W. NAEGELE, and J. J. ELLIOTT: J. Polymer Sci. Pt. B *3*, 729 (1965).
458. DURRELL, W. S., E. C. STUMP, Jr., and P. D. SCHUMAN: J. Polymer Sci. Pt. B *3*, 831 (1965).
459. DONATI, M., G. PEREGO, and M. FARINA: Makromol. Chem. *85*, 301 (1965).
460. KRAUSE, S., J. J. GORMLEY, N. ROMAN, J. A. SHETTER, and W. H. WATANABE: J. Polymer Sci. Pt. A *3*, 3573 (1965).
461. LAL, J.: J. Polymer Sci. Pt. B *3*, 969 (1965).
462. MIYAKE, H., K. HAYASHI, and S. OKAMURA: J. Polymer Sci. Pt. A *3*, 2731 (1965).
463. BAER, E., and J. L. KARDOS: J. Polymer Sci. Pt. A *3*, 2827 (1965).
464. DURRELL, W. S., G. WESTMORELAND, and M. G. MOSHONAS: J. Polymer Sci. Pt. A *3*, 2975 (1965).
465. LUNK, H. E., and E. A. YOUNGMAN: J. Polymer Sci. Pt. A *3*, 2983 (1965).
466. LYMAN, D. J., J. HELLER, and M. BARLOW: Makromol. Chem. *84*, 64 (1965).

467. MANGARAJ, D., S. PATRA, P. C. ROY, and S. K. BHATNAGAR: Makromol. Chem. *84*, 225 (1965).
468. ALLCOCK, H. R., and R. L. KUGEL: J. Am. Chem. Soc. *87*, 4216 (1965).
469. TRICK, G. S.: J. Appl. Polymer Sci. *3*, 253 (1960).
470. NATTA, G., L. PORRI, and G. SOVARZI: European Polymer J. *1*, 81 (1965).
471. MEYER, K. H., W. LOTMAR, and G. W. PANKOW: Helv. Chim. Acta *19*, 930 (1936).
472. SINNOT, K. M.: SPE (Soc. Plastics Engrs.) Trans. *2*, 65 (1962).
473. MILLER, R. L.: Crystallographic Data for Various Polymers. Durham: Chemstrand Research Center, Inc., N. C., 1963.
474. BROWN, C. J., and A. C. FARTHING: J. Chem. Soc. 3270 (1953).
475. NOETHER, H. D.: J. Polymer Sci. *25*, 217 (1957).
476. MORAGLIO, G., G. POLIZZOTTI, and F. DANUSSO: European Polymer J. *1*, 183 (1965).
477. MANGARAJ, D., S. PATRA, and P. C. ROY: Makromol. Chem. *81*, 173 (1965).
478. KERN, R. J.: Makromol Chem. *81*, 261 (1965).
479. SIANESI, D., and G. CAPORICCIO: Makromol. Chem. *81*, 264 (1965).
480. KARASZ, F. E., H. E. BAIR, and J. M. O'REILLY: J. Phys. Chem. *69*, 2657 (1965).
481. ABU-ISA, I., and M. DOLE: J. Phys. Chem. *69*, 2668 (1965).
482. POWELL, J. A., and R. K. GRAHAM: J. Polymer Sci. Pt. A *3*, 3451 (1965).
483. CORRADINI, P.: Atti Accad. Nazl. Lincei, Rend. Classe Sci. Fis. Mat. Nati. *25*, 517 (1958).
484. BADAMI, D. V.: Polymer *1*, 273 (1960).
485. BOILEAU, S., J. COSTE, J. RAYNAL, and P. SIGWALT: Compt. Rend. *254*, 2774 (1962).
486. KAIL, J. A. E.: Polymer *6*, 535 (1965).
487. IMADA, K., T. MIYAKAWA, Y. CHATANI, H. TADOKORO, and S. MURAHASHI: Makromol. Chem. *83*, 113 (1965).
488. CESARI, M., G. PEREGO, and A. MAZZEI: Makromol. Chem. *83*, 196 (1965).
489. NATTA, G., P. LONGI, and V. NORDIO: Makromol. Chem. *83*, 161 (1965).
490. SAOTOME, K., and Y. KODAIRA: Makromol. Chem. *82*, 41 (1965).
491. MATZNER, M., R. BARCLAY, Jr., and C. N. MERRIAM: J. Appl. Polymer Sci. *9*, 3337 (1965).
492. CARAZZOLO, G., L. MORTILLARO, L. CREDALI, and S. BEZZI: J. Polymer Sci. Pt. B *3*, 997 (1965).
493. MANTELL, G. J., D. RANKIN, and F. R. GALIANO: J. Appl. Polymer Sci. *9*, 3625 (1965).
494. COIRO, V. M., P. DE SANTIS, L. MAZZARELLA, and L. PICOZZI: J. Polymer Sci. Pt. A *3*, 4001 (1965).
495. MIDDLETON, W. J., H. W. JACOBSON, R. E. PUTNAM, H. C. WALTER, D. G. PYE, and W. H. SHARKEY: J. Polymer Sci. Pt. A *3*, 4115 (1965).
496. TUBBS, R. K.: J. Polymer Sci. Pt. A *3*, 4181 (1965).
497. RAMIAH, M. V., and D. A. I. GORING: J. Polymer Sci. Pt. C *11*, 27 (1965).
498. DUDEK, T. J., and J. J. LOHR: J. Appl. Polymer Sci. *9*, 3795 (1965).
499. OVERBERGER, C. G., and H. JABLONER: J. Polymer Sci. *55*, S32 (1961).
500. STENSTROM, J. A., and W. F. HALE: J. Polymer Sci. Pt. A *3*, 3843 (1965).
501. CARAZZOLO, G., and G. VALLE: Makromol. Chem. *90*, 66 (1966).
502. Chem. Eng. News 57, Dec. 7, 1964.
503. STAFFIN, G. D., and C. C. PRICE: J. Am. Chem. Soc. *82*, 3632 (1960).
504. TURNER JONES, A.: Polymer *7*, 23 (1966).
505. TURNER JONES, A., J. M. AIZLEWOOD, and D. R. BECKETT: Makromol. Chem. *75*, 134 (1964).
506. BARLOW, M.: J. Polymer Sci. Pt. A-2, *4*, 121 (1966).
507. PRESTON, J.: J. Polymer Sci. Pt. A-1, *4*, 529 (1966).
508. KENNEY, J. F., and G. W. WILLCOCKSON: J. Polymer Sci. Pt. A-1, *4*, 679 (1966).
509. RICE, L. M., and J. B. CLEMENTS (to Celanese Corp.): U. S. 3,161,619 (1964).
510. ALLCOCK, H. R., R. L. KUGEL, and K. J. VALAN: J. Inorg. Chem. *5*, 1709 (1966).
511. ALLCOCK, H. R., and R. L. KUGEL: J. Inorg. Chem. *5*, 1716 (1966).
512. McKINNON, A. J., and A. V. TOBOLSKY: J. Phys. Chem. *70*, 1453 (1966).
513. BECKER, G. W.: Kolloid-Z. *140*, 1 (1955).
514. HEYDEMANN, P., and H. D. GUICKING: Kolloid-Z. *193*, 16 (1963).
515. OKAMURA, K.: Master's Thesis, Syracuse Univ., 1965.

516. THOMPSON, E. V.: J. Polymer Sci. Pt. A-2, *4*, 199 (1966).
517. STROUPE, J. D., and R. E. HUGHES: J. Am. Chem. Soc. *80*, 2341 (1958).
518. KATAOKA, T., and S. UEDA: J. Polymer Sci. Pt. B *4*, 317 (1966).
519. LANDO, J. B., H. G. OLF, and A. PETERLIN: J. Polymer Sci. Pt. A-1, *4*, 941 (1966).
520. SUGAI, S., K. KAMASHIMA, S. MAKINO, and J. NOGUCHI: J. Polymer Sci. Pt. A-2, *4*, 183 (1966).
521. GIBB, T. B. Jr., R. A. CLENDINNING, and W. D. NIEGISCH: J. Polymer Sci. Pt. A-1, *4*, 917 (1966).
522. MARK, J. E., R. A. WESSLING, and R. E. HUGHES: J. Phys. Chem. *70*, 1895 (1966).
523. WESSLING. R. A., J. E. MARK, and R. E. HUGHES: J. Phys. Chem. *70*, 1909 (1966).
524. CESARI, M., G. PEREGO, and W. MARCONI: Makromol. Chem. *94*, 194 (1966).
525. HUGUET, M. G.: Makromol. Chem. *94*, 205 (1966).
526. SHAMBELAN, C.: Dissertation Abstr. *20*, 120 (1959).
527. SAOTOME, K., and K. SATO: J. Polymer Sci. Pt. A-1, *4*, 1303 (1966).
528. OGATA, N., T. ASAHARA, and S. TOHYAMA: J. Polymer Sci. Pt. A-1, *4*, 1359 (1966).
529. SAOTOME, K., and H. KOMOTO: J. Polymer Sci. Pt. A-1, *4*, 1463 (1966).
530. SAOTOME, K., and H. KOMOTO: J. Polymer Sci. Pt. A-1, *4*, 1475 (1966).
531. COLLINS, E. A., and L. A. CHANDLER: Rubber Chem. Tech. *39*, 193 (1966).
532. WETTON, R. E., and G. ALLEN: Polymer *7*, 331 (1966).
533. PRICE, F. P., and R. W. KILB: J. Polymer Sci. *57*, 395 (1962).
534. NAKAJIMA, A., H. HAMADA, and S. HAYASHI: Makromol. Chem. *95*, 40 (1966).
535. IWAKURA, Y., K. HAYASHI, S. KANG, and K. INAGAKI: Makromol. Chem. *95*, 205 (1966).
536. AUBREY, D. W., and A. BARNATT: J. Polymer Sci. Pt. A-1, *4*, 1709 (1966).
537. SMITH, J. G., C. J. KIBLER, and B. J. SUBLETT: J. Polymer Sci. Pt. A-1, *4*, 1851 (1966).
538. BORR, J. Jr., and E. A. YOUNGMAN: J. Polymer Sci. Pt. A-1, *4*, 1861 (1966).
539. NATTA, G., and M. PEGORARO: Atti Accad. Nazl. Lincei, Rend. Classe Sci. Fis. Mat. Nati *34*, 110 (1963).
540. HOBIN, T. P.: Polymer *7*, 367 (1966).
541. BEAUMONT, R. H., B. CLEGG, G. GEE, J. B. M. HERBERT, D. J. MARKS, R. C. ROBERTS, and D. SIMS: Polymer *7*, 401 (1966).
542. HSU, N. N-C.: PhD Thesis, Akron Univ., 1966.
543. REIMSCHUESSEL, H. K.: J. Polymer Sci. Pt. B *4*, 953 (1966).
544. FREDERICKS, R. J., T. H. DOYNE, and R. S. SPRAGUE: J. Polymer Sci. Pt. A-2, *4*, 899 (1966).
545. OTA, T., M. YAMASHITA, O. YOSHIZAKI, and E. NAGAI: J. Polymer Sci. Pt. A-2, *4*, 959 (1966).
546. WILLIAMS, G.: Trans. Far. Soc. *59*, 1397 (1963).
547. FROSINI, V., P. MAGAGNINI, E. BUTTA, and M. BACCAREDDA: Kolloid-Z. u. Z. Polymere *213*, 115 (1966).
548. PEREPELKIN, A. N., and P. V. KOZLOV: Vysokomol. Soed. *8*, 56 (1966); Polymer Sci. U.S.S.R. *8*, 57 (1966).
549. NATTA, G.: Chem. and Ind. (London), 1520 (1957).
550. BACCAREDDA, M., and E. BUTTA: J. Polymer Sci. *51*, S39 (1961).
551. CARAZZOLO, G., and M. MAMMI: Makromol. Chem. *100*, 28 (1967).
552. JONES, L. D., and F. E. KARASZ: J. Polymer Sci. Pt. B *4*, 803 (1966).
553. HELLWEGE, K. H., J. HENNIG, and W. KNAPPE: Kolloid-Z. *186*, 29 (1962).
554. BONART, R.: Makromol. Chem. *92*, 149 (1966).
555. GALPERIN, Y. L., Y. V. STROGALIN, and M. P. MLENIK: Vysokomol. Soed. *7*, 933 (1965); Polymer Sci. U.S.S.R. *7*, 1031 (1966).
556. BACCAREDDA, M., E. BUTTA, V. FROSINI, and P. L. MAGAGNINI: J. Polymer Sci. Pt. A-2, *4*, 789 (1966).
557. PELLON, J.: J. Polymer Sci. Pt. A *1*, 3561 (1963).
558. SAOTOME, K., and H. KOMOTO: J. Polymer Sci. Pt. A-1, *5*, 107 (1967).
559. SAOTOME, K., and H. KOMOTO: J. Polymer Sci. Pt. A-1, *5*, 119 (1967).
560. STILLE, J. K., and J. A. EMPEN: J. Polymer Sci. Pt. A-1, *5*, 273 (1967).

561. GREENE, J. L., Jr., E. L. HUFFMAN, R. E. BURKS, Jr., W. C. SHEEHAN, and I. A. WOLFF: J. Polymer Sci. Pt. A-1, *5,* 391 (1967).
562. FULLER, C. S., C. J. FROSCH, and N. R. PAPE: J. Am. Chem. Soc. *62,* 1905 (1940).
563. BUNN, C. W., and E. V. GARNER: Proc. Roy. Soc. (London) *A 189,* 39 (1947).
564. BEAMAN, R., and F. CRAMER: J. Polymer Sci. *21,* 223 (1956).
565. WOLF, K. A.: Private Communication.
566. MARTUSCELLI, E., R. GALLO, and G. PAIARO: Makromol. Chem. *103,* 295 (1967).
567. HELMUTH, E., and B. WUNDERLICH: J. Appl. Phys. *36,* 3039 (1965).
568. MANDELKERN, L.: Rubber Chem. Tech. *32,* 1392 (1959).
569. WUNDERLICH, B., and M. DOLE: J. Polymer Sci. *24,* 201 (1957).
570. INOUE, M.: J. Polymer Sci. Pt. A *1,* 2697 (1963).
571. JAFFE, M., and B. WUNDERLICH: Kolloid-Z. u. Z. Polymere *216—217,* 203 (1967).
572. WOOD, L. A., and L. W. TILTON: J. Res. Natl. Bur. Std. *43,* 57 (1949).

Offsetdruck: Julius Beltz, Weinheim/Bergstr.

SPRINGER-VERLAG
BERLIN·HEIDELBERG·NEW YORK

Chemie, Physik und Technologie der Kunststoffe in Einzeldarstellungen

Glasfaserverstärkte Kunststoffe